Pittsburgh-Konstanz Series in the Philosophy
and History of Science

Science, Values, and Objectivity

EDITED BY   Peter Machamer and Gereon Wolters

University of Pittsburgh Press / Universitätsverlag Konstanz

Published in the U.S.A. by the University of Pittsburgh
Press, Pittsburgh, Pa., 15260
Published in Germany by Universitätsverlag Konstanz
GMBH

Copyright © 2004, University of Pittsburgh Press

All rights reserved
Manufactured in the United States of America
Printed on acid-free paper

10 9 8 7 6 5 4 3 2 1

Library of Congress Cataloging-in-Publication Data
Science, values, and objectivity / edited by Peter
Machamer and Gereon Wolters.
   p. cm. — (Pittsburgh-Konstanz series in the
philosophy and history of science)
Includes index.
ISBN 0-8229-4237-2 (hardcover : alk. paper)
1. Science—Philosophy. I. Machamer, Peter K.
II. Wolters, Gereon. III. Series.
Q175.S3626 2004
501—dc22
2004013611
A CIP catalogue record for this book is available from
the British Library.

Contents

Preface — vii

Introduction
Science, Values, and Objectivity — 1
Peter Machamer and Gereon Wolters

1 The Epistemic, the Cognitive, and the Social — 14
   Larry Laudan

2 Is There a Significant Distinction between Cognitive and Social Values? — 24
   Hugh Lacey

3 Epistemic and Nonepistemic Values in Science — 52
   Mauro Dorato

4 The Social in the Epistemic — 78
   Peter Machamer and Lisa Osbeck

5 Transcending the Discourse of Social Influences — 90
   Barry Barnes

6 Between Science and Values — 112
   Peter Weingart

7 How Values Can Be Good for Science — 127
   Helen E. Longino

8 "Social" Objectivity and the Objectivity of Value 143
Tara Smith

9 On the Objectivity of Facts, Beliefs, and Values 172
Wolfgang Spohn

10 A Case Study in Objectifying Values in Science 190
Mark A. Bedau

11 Border Skirmishes between Science and Policy
Autonomy, Responsibility, and Values 220
Heather E. Douglas

12 The Prescribed and Proscribed Values in Science Policy 245
Sandra D. Mitchell

13 Bioethics
Its Foundation and Application in Political Decision Making 256
Felix Thiele

14 Knowledge and Control
On the Bearing of Epistemic Values in Applied Science 275
Martin Carrier

15 Law and Science 294
Eric Hilgendorf

Index 311

Preface

These essays were presented at the Sixth Pittsburgh-Konstanz Colloquium, held in October 2002, and prepared especially for inclusion in this volume. The colloquium was co-organized by the Center for the Philosophy of Science at the University of Pittsburgh and the Zentrum Philosophie und Wissenschaftstheorie at the University of Konstanz. The conference was supported by funds from the Harvey and Leslie Wagner Endowment (University of Pittsburgh) and the Fritz Thyssen Foundation (Cologne).

We wish to thank all the participants in the Sixth Pittsburgh-Konstanz Colloquium. We are most grateful to Erika Fraiss (in Konstanz) for the administrative work she has done. We also thank the staff of the Center for Philosophy of Science, Karen Kovalchick, Joyce McDonald, Carol Weber, and Jim Bogen, and the Center's director, James Lennox. And we thank Dennis Pozega for the copyediting he did on the manuscripts.

Introduction

Science, Values, and Objectivity

Peter Machamer and Gereon Wolters

Most people today agree that values enter into science — some values, somehow, somewhere. Few people, if any, still uphold the notion that science in all its aspects is a value-free endeavor. Two places where it is easy to see how values may enter come quickly to mind. First, values affect researchers' decisions about which projects and problems they will work on. That someone chooses to go into AIDS research because it is well funded or because the work may help solve a pressing social problem are clearly cases where values come into play. Some scholars chose research topics that are favored by their dissertation directors. So reasons for problem selections (conscious, well thought out, unconscious, precipitous, however arrived at) often, if not always, contain a value premise.

Another area that values clearly affect comes at the other end of the process of science: the uses to which some scientific results are put. A well-known example is the research in atomic energy that was used to produce the atomic bomb for the purpose of ending World War II. Indeed, problem choice and use of results for project-oriented research for government and corporate research institutions often combine both aspects. Many projects are chosen for the putative utility of their results — for profit, public welfare, or whatever. It is, though, important to recognize that the entry of values into these decisions is not in itself *bad*. Values entering science in these places are only as good or bad as the values themselves.

Now, many things might be said about the nature of values. But let us just keep it simple. We are firm advocates of the "KISS" method of philosophy (keep it simple, stupid). As a first approximation we might say that values are attached to "things" (objects, people, states of affairs, institutions, and so on) that some person, group, or institution believe to be important or significant. In a scientific context, the major way values show their importance is how they enter as premises or bases for making decisions or performing actions in the context of doing science and scientific research. That is, if scientists let certain factors affect and guide their intellectual and practical endeavors, then these factors are what they take to be important (for whatever reason). The various beliefs, techniques, and practices that scientists use to make judgments and evaluations are the loci in which values display themselves. Whether or not scientists can justify these values is a different question, which we will touch on briefly later. We are assuming that values as they occur, at least in this context, can be elucidated and made explicit, for they must play a substantive role in thinking or acting.

What we need to know now is what kinds of values there are, so that we may identify them when they occur, or elicit them when they play some nonpatent role. It also would be nice to know where, other than at the beginning and the end, values are most likely to enter into scientific decision making and practice.

But as soon as we turn to inquiring into the kinds of values there are, a veritable plethora of confusions and unclarities greets us. Just think of cognitive, epistemic, truth preserving, social, cultural, political, emotional, personal, individual, subjective, economic, moral, ethical, religious, aesthetic, prudential, pragmatic, utilitarian, deontological, peer group, and the notorious family values. The mind begins to boggle at the task of getting these all straight and of knowing, for any concrete value uncovered, which kind it is. Consider, for example, the value of having adequate evidence to support your claims. If anything, this seems clearly a cognitive or epistemic value, a value that promotes discovering truth. This value somehow distinguishes science from pseudo-science. Yet, as we shall see in detail in some of the chapters in this volume, when it comes to treating evidence statistically or to selecting the form in which the evidence is presented to test a hypothesis in question, there may be noncognitive aspects.

Simply illustrated, a person is not born with a statistical package that may be used to process data; one must be trained by social and peer group interactions that teach one to value statistical rigor and how to use specific statistical techniques. This knowledge is, in this sense at least, a function of social value. But worse, if one then proceeds by habit to use the statistical reasoning that she has been trained in, without reflecting on its adequacy for the particular project at hand, then in what sense is this an epistemic or truth-seeking decision? At its worst the decision reflects sloth, which is, as is well known, one of the seven deadly sins. Yet, dare we say that this is the way many of us have chosen which statistical package to use? Of course, we might well favor the technique that we feel will present the data in the best light, that is, the one we believe is most likely to show a favorable result for our hypothesis. Again, pure reason as guiding principle is missing or severely compromised. But even if we have critically reflected and come to an honest decision that this technique is the best one for our type of question, many kinds of factors go into this decision. Moreover, we still may implement this value well, sloppily, carelessly, or pedantically. Are there not evaluations hidden in these adverbial descriptions? The multidetermination of such choices is unavoidable, and the choices, and the reasons entering into making them, can, collectively and individually, be either good or bad.

The values that are taken as norms of science and theorizing further exacerbate the problem, for it is not clear how to typify them. Take, for example, the claim in cognitive science that theories ought to be computable. Computability is put forward not only as a desideratum but also as a necessary condition of adequacy. Is it cognitive, aesthetic, or pragmatic? One might want to answer that it is all of the these.

The analyses in this volume help us to identify the crucial values that play a role in science and maybe help us to sort out some of the criteria we can use to do so. Finally, it would be good to be able to elaborate the conditions for warranting certain values as necessary or central to the doing of good science. But maybe this is just vainglorious hope, and we should rest content with being able to clarify the multitude of values and value types that affect our doing of science.

Let us briefly shift attention to the doing of science. We have spoken of two places where nonepistemic values clearly enter science: in choosing research problems and in using results. But what about the role values play in the doing of science itself, in the very activity of scientific research? Certainly, there is much more to science than just research. Decision making in science comes at many different points and in many different ways. While the presence of epistemic values in research is rarely contested, it would seem most difficult to show the influence of nonepistemic values in this area. So our reasoning is, If we can show how nonepistemic values play a role in a research context, it should be easy to show that values play the same or similar roles in any other aspects of the scientific endeavor. Further, it seems easier to show how nonepistemic values occur in the context of discovery, but can we pinpoint where they show up, either as premises in inferences or in necessary presuppositions, in the context of justification?

How can we put some structure on this inquiry? Where in scientific research should we look, and how could we characterize scientific research that would direct attention to the fruitful places? Here is a strategy that has had some moderate success. Why not look at the "stages" of research as they are exhibited in the presentation of that research, namely, at the structure of research as exhibited in a research report or scientific paper. This puts us squarely in the context of justification. If we wished, we could relate each of these "demonstration" steps to the activities that led to the presentation of the research. In this case we could use the schema to move back to the context of discovery. Though, of course, discovery, with its feedback and bootstrapping operations, is never the linear process that most published research papers depict. Here are the six sections that scientific research papers typically have:

1. Choice of research project, problem statement, and review of previous literature.
2. Choice of "experimental" paradigm, taking into account technological availability, as well as background knowledge and conditions.
3. Implementation of experiments (or gathering of data); results of running the experiments.
4. Interpretations of data and results.
5. Discussion section; reflecting on importance of results and integrating them into the existing literature.
6. Implications or uses of results (not often explicitly stated in research papers, except when researcher is seeking future funding).

Many who deny that science can (or should) be value free have focused on the first and last stages of this process, where the intrusion of seemingly extra scientific values is fairly obvious.

An alternative to this model for inquiring into where values enter would be to look at the corresponding stages of developing, running, and completing a scientific research project. This would yield a relatively isomorphic set of stages but might focus attention more on the way values play a role in how a project is conceived and carried out (rather than trying to find their traces in a completed research report).

We have claimed that values enter into science, and particularly into scientific reasoning, in many places. Perhaps the easiest way to see this, and see it in a general way, is to think about the role of classification or categorization in judgment in general, and in scientific judgments in particular. Briefly, and in linguistic terms, to make a judgment is to attach a predicate to a subject. In the simplest cases, this is to make a claim about an object (the subject) that it truly exhibits the property ascribed by the predicate. In other words, to make a judgment is to classify or categorize a subject in terms of a predicate and say that it belongs to that class or category of things.

So in science to sort data into types, or search for data that relate to the theory being tested, is already an attempt to apply categories to particular things. Now, in a trivial way this categorizing is a social activity — people need to be trained in such practices as how to look for things under microscopes. In this way all categorization is just an instance of the fact that all cognitive or epistemic activity involves learning or training, and training is public and social in nature.

More interestingly, many times categorization involves values that are related to issues that are only indirectly attached to the judgment being made. For example, consider the determination of "brain death." Some of the major criteria for determining whether a person is brain dead are profound coma, no corneal reflexes, no cough or gag reflex, and no spontaneous respiratory attempts off ventilator for three minutes (American Medical Association and American Bar Association). This is in contrast, of course, to the older cardiorespiratory criteria of death. Brain death as the criterion for death first came to prominence in 1968, when it was put forward (in slightly different form) by an ad hoc committee of Harvard Medical School (Report 1968). The intentions of those on the committee were clear, that the function of this definition of death, versus the older definition, was to allow for organ harvesting, and so for the possibility of transplants. This is indicated by the guidelines about spontaneous respiratory attempts off ventilator — they read: "Note: If the patient is a potential organ donor, this test should not be done, or if done, should follow the protocol in ICU using 100% oxygen and should be done only after consultation with the Transplant Coordinator."

Notice that whether or not this definition of death is good, fair, or adequate is a different question, demanding another set of arguments. In fact the Germans (in their "Guidelines of the Federal Chamber of Physicians") require that a claim of brain death must involve an irreversible condition that, in the case of primary brain damage, is reassessed and confirmed not earlier that twelve hours after their first pronouncement. In the case of secondary brain damage the reconfirmation must not be earlier that three days after the first assessment. Obviously, this could result in a delay in the harvesting of organs. But here we only point out that accepting brain death as a criterion resulted in a change in the way that persons are classified as dead that was motivated by concerns that incorporate values, namely, that death should allow for the harvesting of organs because organ transplants are valuable.

Another type of case came from Dr. Eric Rodriquez of the University of Pittsburgh Medical School's Geriatric Program. The case was presented in ethical grand rounds to second-year medical students, and it involved a seventy-eight-year-old woman who had fallen, become dehydrated, and was brought to the hospital by ambulance. Her daughter, after being notified, was concerned and distraught. She went on at length about how she worried about her mother, did not know how she could care for her, had two sick children, and so on. In short, the daughter was an emotional wreck. The medical summary for this case was that the mother was capable of being treated adequately as an outpatient but could not live alone in her own apartment.

Students were asked to pretend that they had to make the decision whether to admit the mother. In each group of medical students, the decisions always broke down the same way: Half said they would admit her, half said they would not. Rodriquez pointed out that this split depended on what the medical student thought was the proper role for a hospital. If the student viewed the hospital as primarily an acute-care facility, then there would be no room for the mother. If the hospital's function was to facilitate a broader concept of health, then the appropriate decision would be to admit the mother in order to give the daughter, the primary care-giver, a day or two to prepare herself and her family for her mother's "visit."

Clearly, this is another example in which values enter into decision making. The role someone assigns to a hospital's function depends on the value this person places on hospitals and how he or she assesses their role in the community. This is a big issue and itself involves many values—resource allocation, types of care provision available, scope of physicians' (and health professionals') responsibility to patients, among others. Such considerations about the role of hospitals, and other categorizations or assumptions similar to this one, directly affect decisions that physicians must make every day.

Most everyone is aware of the potential conflict of interest between the needs of a corporation, with its profit motive, and the needs of society to regulate corporations in the interest of public health. When examined closely, the arguments researchers make in their published findings are a mixture of stronger and weaker arguments that point to a conclusion coinciding with their goals as inferred from their place of employment. But this raises the interesting questions of when are bad arguments just bad arguments, that is, the result of faulty reasoning, when they are caused by more complex factors that in their sum constitute bad science, and when are they due to the intrusion of alternative goal-shaping values into the reasoning process. We have tried to show that values have many routes of entry in scientific research, and their presence alone does not make for good or bad science. If it is true that values must enter into scientific decision making and practice, and that the science itself and the character and adequacy of the results depend on these values, then it would behoove us, scientists and philosophers alike, to examine these values. So here is the call: Elucidate the value presuppositions. And, more importantly, be ready to critically assess them.

Values may be decried or defended, blamed or praised. But in a world that needs to strive for a modicum of reason, we urge that values must be critically examined. Values, at least of the kind discussed here, belong to the social world of people acting, so they must be able to be articulated and fit into the space of reasons; they must be articulable as premises in drawing inferences and conclusions. If this is right, critically assessing the values that play a role in scientific research is as crucial to doing good science as interpreting data. In fact, these processes are sometimes the same.

Objectivity

One of the reasons why people are so concerned about the role values play in science is their fear that the objectivity of science, which putatively gives it its unique epistemic status, would be compromised if noncognitive values entered into the actual doing of science. As is well known, modern philosophy of science begins with logical empiricism. One of the axioms of logical empiricism is its noncognitivism, that is, the conviction that value judgments do not make assertions and, therefore, are neither true nor false. Consequently, there can be no knowledge of or truth about values. For logical empiricists, this in turn means that any value judgment, and in general any ethical norm, that pretends to say something true is just pernicious metaphysical gibberish. Value judgments transcend the field of possible rational discourse. Nobody made this clearer than Rudolf Carnap in his grand essay of 1931: "The (pseudo) statements of metaphysics do not serve for the *description of states of affaires*, neither existing ones (in that case they would be true statements) or non-existing ones (in that case they would be at least false statements). They serve for the *expression of the general attitude of a person towards life* ("Lebenseinstellung, Lebensgefühl"). . . . Metaphysicians are musicians without musical ability" (Carnap 1931).

This leaves us with a dilemma (a "value dilemma"): real science from its beginnings (choice of research topics) to its practical uses is replete with value evaluations, and these accrue in such a way that they cannot be eliminated for reasons of principle. Yet science ought to be objective, that is, independent of the personal or group preferences and idiosyncrasies (values included) of the people who conduct it. To put the value dilemma differently, on the one hand, in our real world the logico-empiricist ideal of value-free science cannot exist, and, on the other, the unavoidable value-ladenness of science, for strictly observing logical empiricists, makes science a basically irrational enterprise, lacking exactly the epistemic virtue, objectivity, that made for its success. This conclusion, as far as we know, has not been drawn by anybody close to logical empiricism, or to scientific philosophy, for that matter. Quite to the contrary, in this camp science is the objective enterprise *par excellence*.

The value dilemma points to a major defect in the overall strategy of logical empiricism: its concept of rationality. This concept is understood in a way that is too ideal and this means too restricted for the real world. It prevents logico-empiricists from doing what seems obligatory in order to save the objectivity of science, that is, reasoning (which above all is something rational!) about the values principally involved in their practice. Therefore, to save the objectivity of science, we must free it from an ideal of rationality modeled after mathematics and logic; we must show that both rationality and objectivity come in degrees and that the task of good science is to increase these degrees as far as possible.

So the real question today is where and how particular values enter into science, and how values of many different kinds enter into the doing of "good" science and, concomitantly, how one may avoid "bad" values and thereby bad science. Focusing on these questions will allow us to consider the nature of rationality, objectivity, and knowledge in a new way so that we can address issues dealing with social *and* epistemic (or cognitive) values, especially how social and empirical normativity relate to normative criteria of evaluation. In passing, we shall remark on how all of these concepts relate to the "deep" nature of objective empirical science as practiced in socialized scientific communities. Much of this account was inspired by John Haugeland (1998).

Norms are like rules or the principles that guide regulated activities. They are embodied in performance skills; they are constitutive of knowing how to do something, be it intellectual, practical, or productive. They apply to individuals or groups. They apply in specific social locations. And they are public: they can be learned and observed and have the possibility of being used to correct actions or practices that do not accord with them. A norm is *constitutive* of a practice's being the kind of practice it is (the kind of game it is, the kind of form of life it is). It is by doing things "in this way," and only this way (though there is often some latitude), that a practice is exhibited or defined. Practices, of course, can change, be revised, be ignored, or be discarded. They exist only in specific locations at specific times. They are historical "objects."

Norms specify for individuals or groups when practices are being carried out correctly or properly. Those engaged in a practice, be it science or wine tasting, must be able to determine when a procedure being followed is incorrect (or illegal). This is necessary for understanding or evaluating correctness, and part of knowing what is correct is knowing which possibilities are excluded as incorrect. For example, we must be able to check a child's understanding of arithmetic or geography, not only that the child can give the right answers but also knows when others are wrong *and* can take steps to correct them. The child also needs to make moves toward correcting his or her own mistakes after making them. We may even demand more of the child, making the child exhibit the protocols by which he or she came up with the answer and discuss why some other approach would have been inferior. The ability to check and correct within some conception of the space of possibilities allowed and disallowed by the normative constraints on a practice is what makes the difference between understanding and mere mechanical or rote performance. (This is one of the points that therapists have noted when requiring *working through* as a condition for experiential as opposed to just intellectual knowledge.)

These conditions on practices may make norms seem wholly social. And it is at this point in the exposition where social constructivists often stop. Yet they are justified in part. Norms are social. But what do we do about the objects and activities of nature that ought to be of concern to any good empiricist? For after all it is responsiveness to nature and nature's objects and their working that provides a, if not *the*, major demarcation criterion by which science is distinguished from other human practices. Any account of science must explain how nature, its objects, and its activities function in situations and practices of testing and how objects provide evidence.

We shall use the term "object" in what follows, for it clearly relates to the term "objective." But by "object" we mean the entities and activities that are part of nature or that are part of the causal structure of the world. Objects, *as we know them*, are conceptualized (maybe even linguistized). Objects are partly what they are because they play a role in the system of concepts we have in memory and in the inferences and expectations we have about them. This is what it means for them to be conceptualized. They act or function in a system (or sets of systems) of conceptual practices. Some of our knowledge consists in this interconnectedness, in the schemata or models that comprise our memory or knowledge "representation" systems. The systems involved here are not just declarative or semantic systems but also many grades and kinds of procedural systems.

Yet objects also figure in systems of physical practices and procedures, and here it may be easily understood how objects are recalcitrant. Sometimes objects do things we do not want, expect, or even know how to handle, intellectually or physically. No one can make a silk purse out of a sow's ear, no matter what practice one is engaged in. This was one of the reasons alchemy failed as science, yet its very possibility is one reason why it persists in the realm of fantasy. Humans cannot do with objects what they will or wish, for the objects constrain our very actions, including the activity of thinking.

If the objects, even as we conceptualize them, fail to perform as we had expected on the basis of our memorial schemata, then we have to step back and figure out what is going on in our normative, public schemata system, and so in our expectations. Our interactions with the world have upset the norms we thought were in place regarding what physical objects would do. In such cases we may change our practices, our expectations, or even our normative systems in which those objects figured. But recalcitrance is not always object driven. Sometimes it is experiential or proprioceptive. One cannot ride a bicycle without experiencing being able to balance or build a tower from children's blocks without a sense that they are being set up correctly. Such feelings are still part of the physical. Acting in the world is one of the major ways of getting involved with the physical. In fact, the physical recalcitrance of objects is not always unexpected. Often, the fact that objects place constraints on our actions is built into the conceptual scheme we have. This is why we do not flaunt the rules of the road while driving. We do not obey the rule not to drive on the wrong side of the highway because we may go to jail (social constraint), but rather because we may get hit by another car. We fear injury and death, outcomes that some kinds of objects can cause in us. In this line of reasoning lies the realists' point that one does not want a social constructivist to design our airplanes.

Yet, as already said, normativity depends on the conceptual and physical roles of those objects and their set of associated practices (including expectations and inference licensing) But another part of normativity depends on objects' recalcitrance, which affects our interactions with those objects. Physical objects carry their own set of constraints on our actions and practices, including the intellectual ones. But the physical and social norms involved here are not incompatible with one another. If they were, then the debate between the "realists" and the "social constructivists" would be right headed. It should be clear that it is necessary to have both.

Maybe this point can be clarified by looking at a perception case, by thinking about *seeing as*. If, for example, I see object X as F; then the F is part of the conceptual, categorical representation I have, and it has associated schemata that are the norms for Fs. But X is an object (mostly spatiotemporally discrete objects), and it has determinate physical properties and activities. If I hallucinate and have an F-like experience where there is no X, then I, and certainly others, may check this. In this view of perception no special priority is given to first person singular experiences or access. There are no criteria for knowledge that are "subjective" or that may be found in the brain or mind of the knower. Yet this does not entail that one's "subjective" experiences in the form of conscious awareness play no role in one's own knowledge We cannot elaborate this claim here, but we saw one instance of it in the case of proprioceptive experience while riding a bicycle.

For me to think about the object, its properties, and activities, I have to conceptualize them, but my conceptualization cannot change the way they act or change how the objects respond to me when I act. Sometimes, too, I cannot change the way I act or the character of my experience because these are determined by my physical system or by the objects in the world.

Perception like knowledge is only in part in the head; the rest inextricably and ineliminably is action, and more basically interaction with a recalcitrant world. When things interact with us in unexpected ways, then we must change our actions (check our observations), concepts, or the schemata that gave us the expectations and that we used to deal with those things we thought were F-like.

In sum, objectivity lies in the objects and how they affect us, whether we understand them or not. The objects constrain our actions and thoughts, though the norms for these thoughts and actions lie in the public, social world. It is the objects that are impartial, and provide the basis for our nonpartial conceptualizations. This is the only kind of objectivity there is. It is sufficient for science if science is conceived as a rational, fallible, human, historical endeavor. If science is not so conceived, then science does not exist.[1]

NOTES

Part of this chapter was presented at a workshop sponsored by the National Science Foundation, "Values in Scientific Research," held at the Center for Philosophy of Science, University of Pittsburgh, October 1998. An early version of the material present here was worked out with Merrilee Salmon, when she and Machamer presented "Values in Science" to the University of Pittsburgh's Department of Medicine, Hematology Section, at Research Integrity Seminar in May 1997. Many of these themes are elaborated on in different ways at greater length in Machamer and Douglas (1998); see also Machamer and Douglas (1999).

1. Many of the themes treated here are also taken up in McGuire and Tuchan-

ska 2000; although they tackle these themes in a very different manner and have different heroes, our approaches have much in common. Parts of them are developed in Machamer and Osbeck (2002, 2003).

REFERENCES

Carnap, Rudolf. 1931. The elimination of metaphysics through logical analysis of language. Reprinted in *Logical positivism*, ed. A. J. Ayer, 78, 81. Glencoe, IL: Free Press, 1959.

Haugeland, John. 1998. Truth and rule following. In *Having thought: Essays in the metaphysics of mind*. Cambridge, MA: Harvard University Press.

Machamer, Peter, and Heather Douglas. 1998. How values are in science. *Critical Quarterly* 40:29–43.

Machamer, Peter, and Heather Douglas. 1999. Science and values: Comments on Lacey. *Science and Education* 8:45–54.

Machamer, Peter, and Lisa Osbeck. 2002. Perception, conception, and the limits of the direct theory. In *The philosophy of Marjorie Grene. The library of living philosophers*, vol. 29, ed. Randall E. Auxier and Lewis Edwin Hahn, 129–46. Lasalle, IL: Open Court.

Machamer, Peter, and Lisa Osbeck. 2003. Scientific normativity as nonepistemic: A hidden Kuhnian legacy. *Social Epistemology* 17:3–11.

McGuire, J. E., and Barbara Tuchanska. 2000. *Science unfettered*. Athens, OH: Ohio University Press.

Report of the ad hoc committee of the Harvard Medical School to examine the definition of brain death. 1968. A definition of an irreversible coma. *Journal of the American Medical Association* 205:337–40.

1

The Epistemic, the Cognitive, and the Social

Larry Laudan
Department of Philosophy, University of Mexico City

My aim in this chapter is to direct attention to a matter philosophers of science barely examine let alone satisfactorily address: the relationship between the philosophy of science and the theory of knowledge. Like many things we take for granted, this relationship is not well understood. Most philosophers, whether of science or the more traditional sort, would respond that the philosophy of science is applied epistemology — that is to say, it brings the categories and tools of analytic epistemology to bear on understanding the practices called science. Sidney Morgenbesser was, I believe, voicing the conventional wisdom when he quipped in 1968 that "philosophy of science is epistemology with scientific examples" (Morgenbesser 1967, xvi).[1] (There is, of course, another aspect to the philosophy of science — traditionally called "foundations" of science — that is seen not as applied epistemology but rather as applied metaphysics, a topic I do not investigate here.)

This probably all seems harmless enough, not least because of its utter familiarity. But this way of construing the provenance of the philosophy of science is not innocuous. To the contrary, the notion that one can make sense of science by conceiving of it principally as epistemology teaching by example is not only hubris on the part of the epistemologist — hubris after all is an occupational hazard of the philosopher and thus forgivable — but also, I will argue, presupposes the correctness of one particular approach within the philosophy of science, specifically epistemic realism, while denying legitimacy to various other philosophies of science that have at least as distinguished a track record as realism itself does. Still worse, the vision of philosophy of science as applied epistemology forces us to treat as irrational many of the most interesting and important evaluative strategies used in sound science.

I argue that the philosophy of science is not, and should not be conceived as, an exclusively or even principally epistemic activity. This is because science is neither exclusively nor principally epistemic. I will arrive at these interrelated conclusions by a slightly circuitous route. I will begin by focusing on a familiar and specific example of the thesis that philosophy of science is reducible to epistemology. I refer to the conception of the philosophy of science as rational reconstruction, especially as that notion was developed by Hans Reichenbach (and, to a lesser degree, by Rudolf Carnap) in the 1930s and 1940s. I will show that Reichenbachian reconstructions are not simply, as some might have supposed, broad-based case studies of philosophically interesting episodes in science but are instead subject to severe constraints with respect to which bits of real science are reconstructible and which are not. I will show that these constraints are imposed by the acute limitations of the tools of epistemology.

The first point to establish, and it is easy work, is that Reichenbach saw rational reconstructions as devices for identifying the epistemically salient features of any given scientific episode. This means that they are — and this is the first important thing to note about them — rational reconstructions only in a very attenuated and idiosyncratic sense. As conceived by Reichenbach in the opening chapter of his *Experience and Prediction*, rational reconstructions are not attempts to clean up the details of a scientific episode by showing how or to what extent the elements of the episode promoted the ends of the investigator. That sort of instrumental rationality is patently not what Reichenbach had in mind when he talked about rational reconstructions. Rather, for him, the freight that the term *rational* carries in that phrase was purely epistemic. He argued that the only features of any actual situation that appropriately belonged in a rational reconstruction were those bearing on the truth or the falsity of the theory or hypothesis being evaluated in the episode in question. I repeat: for Reichenbach, rational reconstructions were purely and simply epistemic reconstructions. Insofar as the actual case involved activities or principles that, however rational in their own right, had nothing demonstrable to do with determining the truth or falsity of an hypothesis, those activities and principles found no rightful place in the so-called rational reconstruction of the case. The same point applies to Reichenbach's oft-mentioned but little understood distinction between the contexts of discovery and justification. This set of polarities marked, for him, not different temporal stages in an investigator's research but simply the difference between a descriptively rich but philosophically irrelevant account of an episode of the sort you might see in a history book and the very different but purely epistemic account that was to constitute, for him, the rational reconstruction of the episode. For Reichenbach the context of justification consisted of all and only those factors essential to the epistemic evaluation of the theory in question. Everything else — that is, everything not epistemic — was relegated to the context of discovery and consigned to the psychologist or the anthropologist for further investigation. The philosopher's interest in the episode was limited strictly to those elements that passed epistemic muster.

Now, if you support the idea that philosophy of science is applied epistemology, you may find nothing unseemly in Reichenbach's delineation of the task of philosophy of science as that of working with rational (understood now as epistemic) reconstructions of episodes rather than the episodes themselves. Besides, you might add, any philosophically interesting account of any human practice will have to simplify and idealize the blooming, buzzing confusion of the real world in order to have a manageable unit of analysis. I have no problems with simplifications, not even with oversimplifications, when they serve a useful purpose. But what is fishy here is that much of what drives scientific activity, even scientific activity at its rational best, are concerns that have no epistemic justification in a strict sense and that must be excluded from rational reconstructions of science as understood by Reichenbach and most of the others who have construed the philosophy of science as applied epistemology. The rest of this chapter will attempt to deliver on this claim. I aim to show that many, and arguably most, of the historically important principles of theory appraisal used by scientists have been, though reasonable and appropriate in their own terms, utterly without epistemic rationale or foundation.

I focus on one family of examples, among many that I might have chosen. My central argument will depend on noting the frequency and persistence with which scientists insist on evaluating theories by asking about their scope and their generality. Several familiar and important rules of thumb in theory appraisal speak to such concerns. For example, acceptable theories are generally expected to explain the known facts in the domain ("saving the phenomena"), explain different kinds of facts (consilience of inductions), explain why their rivals were successful (the Sellars-Putnam rule), and capture their rivals as limiting cases (the Boyd-Putnam rule). I trust no reader will dispute the ubiquity of rules of this sort in evaluating scientific theories. The question is whether such rules have, or can be conceived to have, any grounding in epistemology per se.

Consider the first rule on this list, to explain the known facts in the domain. Steady-state cosmology was rejected in the 1960s not because it had been refuted but because it offered no account of the cosmic background radiation discovered at Bell Labs. The uniformitarian theories of Hutton, Playfair, and Lyell were rejected by most nineteenth-century geologists, not because they faced massive refutations, but because they steadfastly refused to say anything about how the earth might have evolved from its primitive initial condition to the condition of habitability. Plate tectonic theory triumphed in the 1960s over stable continent geologies principally because the former, but not the latter, could explain long-familiar patterns of continental fit and similarities of fauna and flora between the Old World and the new. Galileo famously argued for the rejection of Ptolemaic astronomy because it could not explain why Jupiter should have moons or why the sun should have spots. He plumped for the Copernican alternative because it could explain such facts about the solar system. The Jovian moons and the sun's spots did not refute any claim in Ptolemy's *Almagest*. Their potency derived from Ptolemy's system evidently lacking any mechanism for making sense of such phenomena.

More generally, it should be uncontroversial that scientists frequently argue for one theory over another if the former can explain or predict something about the world not accounted for by its rivals. I daresay no one regards this as a specious form of argument against a theory. Few would quarrel either with the notion that a theory is, all else being equal, better if it can explain or predict facts from different domains or if it can show its rivals to be limiting cases. This is, nonetheless, a form of argument that has, and can have, no epistemic foundation. Our other three rules about the scope of a theory exhibit the same disconnection from epistemology.

None of these rules can have an epistemic rationale since it is neither necessary nor sufficient for the truth of a statement that it exhibit any of these attributes. That a statement fails to explain a fact with which it is strictly compatible is no argument against its being true. Indeed, most true statements do not exhibit this virtue. Similarly, the fact that a statement explains only one type of fact, rather than several, is no reason to believe that it is false. Indeed, most true statements do not explain different kinds of facts. Likewise, the fact that one statement cannot explain why one of its contraries worked so well is no argument against its truth since most true statements cannot explain why their contraries, if successful, are successful. Finally, we do not generally expect true statements to be such that some of their contraries can be shown to be limiting cases of them.

If these attributes of scope and generality are virtues, and I believe they are, they are not epistemic virtues. They address questions about the breadth and range of our theories rather than questions about their truth or probability. (It is true that philosophers have sometimes tried to describe these virtues of scope as epistemic virtues. Recall, for instance, William Whewell's labored but unsuccessful efforts to show that consilience-achieving inductions are bound to be true. Boyd and Putnam tried to argue that capturing a predecessor as a limiting case was an argument for the truth of a theory.) As I have argued elsewhere (Lauden 1981), no one has shown that any of these rules is more likely to pick out true theories than false ones. It follows that none of these rules is epistemic in character.

Indeed, one can piece together a perfectly general proof that these attributes cannot be epistemic indicators. I do not set much store by such arguments myself, but for those who do, they look like this:

Let T be some theory exhibiting one of the virtues of scope, v. Now, T will have many consequences; and infinitely many of those consequences will lack v, since many of the logical consequences of a statement of broad scope will fail to exhibit such scope. Focus on any one of those consequences, which we will call c_1. Now, if T is true, c_1 must likewise be true. If T is highly probable or verisimilar, c_1 must be even more probable or have more verisimilitude. In short, thanks to the truth-preserving character of entailment, c_1 will necessarily possess all the epistemic virtues of T, while failing to have v. It follows that v cannot be an epistemic virtue since statements failing to exhibit v (such as c_1) are at least as solid epistemically as statements like T, that exemplify v.

This fact should discomfort no one save the epistemologist. It does not show that subjective values drive science or that "merely aesthetic" yardsticks predominate. What it does show is that scientists have expectations about good theories that go well beyond worries about their veracity. If you have any residual doubts on this score, simply ask yourself whether any serious scientist would countenance every statement that he or she believed to be true to be an acceptable theory. Scientists may regard truth as an important virtue (we can argue about that another time). But what cannot be gainsaid is that there are other virtues of theories that loom at least as large in theory evaluation as truth does. By definition, these cannot be epistemic virtues since many false statements exhibit them and many true statements do not. By definition, they can find no place in a so-called rational reconstruction of science driven by an epistemic agenda.

Bas van Fraassen famously argued that a theory does not have to be true to be good. We can add to that dictum a new twist: a theory does not have to be false to be bad. A theory may be bad because it fails the test of possessing the relevant nonepistemic virtues. In other words, we expect our theories to do much work for us, work of a sort that most merely true statements fail to do. However, we may cash out precisely what that additional work is, and when we do so, we will move beyond the epistemic realm into one I call cognitive but nonepistemic values.

Such values are constitutive of science in the sense that we cannot conceive of a functioning science without them, even though they fail to be intelligible in the terms of the classical theory of knowledge. These values have nothing to do with philosophical semantics or with justification conditions, as usually understood. For that reason, I call them cognitive virtues or values, of which the epistemic virtues form a proper (I almost said uninteresting) subset. I have focused thus far on one family of cognitive virtues having to do with the range or scope of theories. Another family of such concerns is whether, in Phillip Kitcher's language, the theories in question achieve "explanatory unification."[2] Like the virtues of scope and generality, the virtue of explanatory unification cannot—counter to claims that Kitcher has sometimes made about it—be a truth-related virtue since it is obvious that every unifying theory, T, will entail non-unifying counterparts, $T_1^*, T_2^*, \ldots, T_n^*$, which must be true if T is true. If scientists regard T as a better theory than any of those weaker counterparts—and they invariably will—this must be because T possesses, and T_i lacks, virtues that are nonepistemic.

If all of this is even half correct, we see that the Reichenbachian formula for putting together a rational reconstruction is fatally restricted and that this restriction speaks to the limits of application and relevancy of epistemology itself. What goes on in science at its best eludes the resources of the theory of knowledge to explain or to justify. Moreover, the Reichenbachian formulation declares to be philosophically irrelevant—mere fodder for the psychologist, the anthropologist, and the sociologist—many of the most important factors that go into theory evaluation in the sciences. Using Reichenbach's own language, my criticism is that he shunts far too much into the context of discovery and leaves little more than bare bones in the context of justification.

If, however, we were to understand rational reconstruction as a technique for analyzing science using the cognitive values that constitute it, not just the epistemic values, the line of demarcation between these two contexts would shift drastically. The context of justification would now recognize concerns about scope, generality, and range of application—and possibly explanatory scope as well—as a part of the rational reconstruction of any episode. The fact that such factors are nonepistemic would be neither here nor there, since rational reconstructions in terms of cognitive values would not, on my proposal, be limited to strictly epistemic factors. Imre Lakatos once argued that the appropriate criterion for evaluating the adequacy of a rational reconstruction of an episode in science involves asking how much of the activity of the scientists involved is captured by the rational reconstruction. By that yardstick, cognitively based reconstructions are clearly preferable to epistemically based ones.

But much more is at stake here than what sort of rational reconstructions we countenance. As I said at the outset, the notion of a rational reconstruction is merely a stalking horse for a larger target of my critique. I refer, of course, to analytic epistemology itself. I submit that once we consider the role that issues of scope, generality, coherence, consilience, and explanatory power play in the evaluation of scientific theories, it becomes clear that science is an activity only marginally or partially epistemic in character. Because that is so, the instinct to reduce the philosophical analysis of science to epistemic terms alone—and there are entire philosophies of science (Bayesianism, for example) committed to doing just that—must be resisted.

Likewise infected by the reduction-to-epistemology bug is the whole of the statistical theory of error. As everyone knows, statisticians recognize only two types of error: accepting a false hypothesis and rejecting a true one. But, of course, once we see that science has aims other than the truth, we recognize that there are whole families of error types associated with each of the relevant cognitive values. Thus, a scientist may accept as explanatory a theory that is not explanatory and reject as nonexplanatory a theory that is. Likewise, a scientist may wrongly believe a theory to be capable of explaining different types of phenomena. These errors — which find no place in the contemporary theory of error — can be just as fatal to a theory as the more familiar errors of accepting the false and rejecting the true. Statisticians do not recognize these errors as errors because they are not epistemic errors. The error statistician, like the Reichenbachian rational reconstructor, takes his cues entirely from the epistemologist. That would be intelligible if and only if we had reason to suppose that the only demand scientists appropriately made of their theories was that they be true. We have no reason for such a premise. Consider, for example, that most scientists would reject, or at least consider badly flawed, a theory that failed to explain prominent facts in its intended domain of application, however good that theory was at capturing the facts it chose to explain. Error statisticians can no more make sense of such an appraisal than epistemologists can. Since scientists do make these additional demands on a theory, and have good reasons for doing so, it is time that the statisticians, like the epistemologists, recognized how severely limited the tools are that they currently bring to the task of explicating scientific rationality. I was sorely tempted to make the same observation about the Bayesians who, like the error statisticians, are fixated on adjusting probability assignments, and indifferent to the role of values in theory assessment other than that of avoiding a Dutch book and falling into losing betting strategies about the truth. I said I was tempted, but I know better, since ingenious Bayesians, like Paleozoic omnivores, can find some way of digesting almost anything in their path.

By way of summary of the argument thus far: We would like to have theories that are true, elusive as that ideal may be. But we would also like theories that are of great generality, that focus on the things we are particularly interested in understanding, that explain as well as predict, and that consolidate existing successes while moving us beyond them. Of all these matters save the first, the epistemologist knows little or nothing. Because, like the early Wittgenstein, epistemologists cannot speak of things they know not; they must maintain a studied silence with respect to most of the values that drive scientific research.

To this point, I have said nothing about the third element in my title, the social. This is perhaps just as well since my thoughts on that topic are more fluid than on the other two. But if forced to fulfill the contract implicit in my title, here is what I would say, at least on Thursdays: There is a century-long philosophical tradition, dating back at least to Marx and Mannheim, of supposing that theories whose acceptance seems to involve exclusively epistemic values do not require the same sort of social-psychological explanation as theories lacking the epistemic virtues. Recall that the whole theory of ideology was developed specifically to explain why people came to believe notions for which there were no compelling epistemic arguments. Strictly speaking, there is something wrong with this way of proceeding since epistemic factors themselves function in, and evolve out of, social interactions among inquirers. In that very broad sense, every human artifact, including human beliefs and conventions about belief authentication, is grounded in social processes of communication, negotiation, and consensus formation. But this sense of the term *social* is so broad as to be vacuous. What I think Reichenbach had in mind, when he identified the context of discovery as the appropriate domain of the social, is the idea that social processes of belief fixation that lack an epistemic rationale are of no philosophic interest (except perhaps as sociopathologies) and their study should be left wholly to the social scientists. By contrast, thought Reichenbach, where there is an epistemic justification for a belief, the philosopher has a legitimate interest in exploring that justification and in arguing the relevance of that justification to the belief itself. If we are minded to draw a line between the social and the rational along these lines, my own suggestion, of course, would be that the philosopher should lay claim to interest in all beliefs for which there is a cognitive rationale, as opposed to only those beliefs for which there is an epistemic rationale. Unlike Mannheim, who defined the scope of the sociology of knowledge in terms of beliefs for which no epistemically compelling rationale exists, I would prefer to see that scope defined in terms of beliefs for which there is no cognitive rationale. Still, as I said earlier, my views on the relationship of the cognitive to the social are too complex, and too tentative, to be reduced to a simple formula.

What I have no hesitation about is my insistence on the explanatory poverty of purely epistemic values and the resultant need to talk philosophically about science in categories that go well beyond the merely epistemic.

NOTES

1. During the late 1960s, Lakatos frequently made similar observations during his seminars at the London School of Economics.
2. Kitcher has formulated this argument in many places; for its most detailed elaboration see Kirchner (1993).

REFERENCES

Kitcher, P. 1993. *Advancement of science*. Oxford: Oxford University Press.
Laudan, L. 1981. A confutation of convergent realism. *Philosophy of science*. 48:19–49.
Morgenbesser, S., ed. 1967. *Philosophy of science today*. New York: Basic Books.
Reichenbach, H. 1938. *Experience and prediction: An analysis of the foundations and the structure of knowledge*. Chicago, IL: University of Chicago Press.

2

Is There a Significant Distinction between Cognitive and Social Values?

Hugh Lacey
Department of Philosophy, Swarthmore College

1. Introduction

There is a distinction between cognitive and social values. But is it significant? Cognitive values are characteristics that scientific theories and hypotheses should possess for the sake of expressing understanding well. Or, as Larry Laudan puts it, they are attributes that "represent properties of theories which we deem to be constitutive of 'good' theories" (Laudan 1984, xii). Social values are characteristics deemed constitutive of a "good" society. Does this distinction illuminate key features of scientific knowledge and practices?

The tradition of modern science would answer this question with a resounding yes, and this underlies the commonly held view of science as "value free." In accordance with this view, social values (or any other kind of noncognitive values) are not among the criteria for a theory's being acceptable. They have nothing to do with appraising the understanding expressed in scientific theories; the judgment to accept a theory is *impartial*. In addition, the broad characteristics of scientific methodology should be determined only in response to an interest in gaining deepened understanding of phenomena, and particular social values should not systematically shape the priorities and direction of research: methodology is *autonomous*. Science holds impartiality and autonomy as ideals, values of scientific practices that may not always be well manifested in actual fact. Usually *neutrality* accompanies them.

Science per se gives privilege to no particular social value. Scientific theories are *cognitively neutral:* social value judgments are not among their entailments; and they are *neutral-in-application:* in principle they can inform evenhandedly the interests of the wide range of social values that can be viably held today. To a first approximation, these views together sum up what is meant by the statement *science is value free.* The integrity, legitimation, prestige, and alleged universal value of science have often been tied to three values—impartiality, autonomy, and neutrality—being highly manifested (with a trajectory of higher manifestation) in the practices of science, for, it is said, science conducted in such practices has enabled the technological applications that have so transformed the world in recent times.[1]

While I reject some of the components of the notion that *science is value free* (Lacey 1999, 2002a), I too think that there is a significant distinction between cognitive and social values. I will argue that the distinction is crucial for properly interpreting the results of scientific research and for opening up reflection on how neutrality might be defended as a value of scientific practices at a time when much of mainstream scientific research is becoming increasingly subordinate to "global" capitalism. My defense of the distinction, unlike the traditional viewpoint, does not support keeping social values out of the core of scientific activity, although I continue to endorse impartiality.

Summary

My argument is complex. In the next section I formally distinguish cognitive and social values and clarify the relationship between factual and value judgments. Drawing on what I take to be the aim of science, I argue in the third section that being of service to interests fostered by any social values cannot be considered a cognitive value. This leaves it open, however, that—as a matter of fact—the aim of science is best served, and so the degree of manifestation of the cognitive values best augmented, by conducting research under a "strategy" that makes it likely that its theoretical products will be especially well-attuned to serve some special interests. Features of modern science may suggest that this is true. In the crucial fourth section I maintain that modern science is conducted almost exclusively under a particular kind of strategy (materialist strategies). In turn, that it is conducted in this way is explained and rationalized by widespread commitment to certain specifically modern ways of valuing the control of natural objects and mutually reinforcing relations between them and the conduct of research under materialist strategies. These relations are so pervasive that it is often not appreciated that for certain domains of phenomena of interest (in agriculture, for example) where the modern ways of valuing control are contested there may be possibilities that cannot be encapsulated under materialist strategies but can be under alternative strategies. I propose agroecology as a salient example. The aim of science does not license ignoring such alternatives. From this I go on to conclude that unless research is conducted under a variety of strategies, the linkage of modern science with the values about control cannot be broken and, because of this, important avenues of research will not be explored. Finally, in the fifth section, I identify three "moments" of scientific activity: adopting a strategy, accepting theories, and applying scientific knowledge. Social values may play legitimate roles at the first and third moments, but at the second moment social values have no legitimate role alongside the cognitive values. The significance of the distinction between cognitive and social values comes from the central place of the second moment, when judgments about what counts as sound scientific knowledge are made. It leaves an important role for social values at the other moments.

My argument frequently makes use of substantive proposals that I have elaborated and defended elsewhere. Obviously I cannot expect the reader to endorse them on my word. My claim that the distinction between cognitive and social values is significant ultimately rests on the explanatory and justificatory roles (defended in works of mine cited throughout the chapter) that it may play. Since I cannot offer more detail in this chapter, it might be best to read it as a proposal that supports an account of the structure of scientific activity that, if well established, would show the significance of the distinction. I know of no simpler way to make the argument.

2. Values, Value Judgments, and Value-Assessing Statements

We hold various kinds of values: personal, moral, social, aesthetic, cognitive; they are held in more or less coherent and ordered *value outlooks* in which they mutually reinforce one another.[2] To illustrate some of the general features of values, let ϕ denote a particular kind of value, v denote some characteristic that (typically) can be *manifested* to a greater or lesser degree in ϕ, and X denote a person. Then X holds v as a ϕ-value, if and only if

1. X desires that v be highly manifested in ϕ;
2. X believes that the high manifestation of v in ϕ is partly constitutive of a "good" ϕ; and
3. X is committed, *ceteris paribus*, to act to enhance or maintain the degree of manifestation of v in ϕ.

In the case of social values, ϕ represents society (social institutions, social structures), and a social value (sv)—such as *respect for human rights*—is something whose high manifestation is valued in society. In the case of cognitive values, ϕ represents accepted theory (hypothesis), and a cognitive value (cv) is something whose high manifestation is valued in accepted theories (hypotheses), that is, in confirmed bearers of understanding.[3]

"X holds the ϕ-value v" will be considered equivalent to "X makes the ϕ-value judgment, V, of valuing that v be a well-manifested feature of ϕ." (V represents the value judgment made when v is held as a value.) Basic value judgments are broadly of three types: "that v is a ϕ-value, a characteristic of a 'good' ϕ"; "that v_1 ranks more highly (as a ϕ-value) than v_2"; and "that ϕ manifests v to a 'sufficiently high' degree." There are also value judgments of the form: "that u has ϕ-value" or "that u is an object of ϕ-value," judgments made on the basis of u's contributing to the manifestation of the ϕ-values.

A methodological rule can be said to have cognitive value through its causal contribution toward generating or confirming theories that manifest cv's highly. On similar grounds, scientific practices organized so that they manifest certain sv's (Longino 1990, 2002) or the cultivation of certain moral virtues among scientists may be proposed to have cognitive value. It is also likely to be an sv for X that the cv's (X holds) be socially embodied: that there be social institutions (scientific ones) that nourish practices in which theories (more and more of them) highly manifesting these cv's come to be accepted; and a theory (T) that manifests cv's highly may also have social value (for X), by virtue of its making a contribution—on application or perhaps just by providing understanding of some phenomena—to the manifestation of X's sv. Thus, "is an object of social value in light of a set of sv's" can designate a property of T. In affirming that cv's and sv's are distinct, I am committed to the claim that there is a set of cognitive values and that "is an object of social value in light of a set of sv's" is not one of them.

There are close links between the values that X holds and those embodied in the institutions in which X participates; and, at least sometimes, that these are embodied in the institutions explains why X holds these values rather than others. For those who consider values to be subjective preferences, explaining the values that X holds is the end of the story. On my account, values are not reducible to subjective preferences. Another distinction needs to be brought into play:[4] that between X's value judgment V that v is a φ-value and a statement about the degree of manifestation of v in φ. To what degree φ manifests v is a matter of fact. I will call statements expressing such matters of fact *value-assessing statements*. They are not value judgments, although statements expressing that v is manifested in φ to a "good enough" degree are; they are hypotheses that can be part of theories investigated in the social sciences and can be appraised using available data and cv's.

There are deep interconnections between value judgments and value-assessing statements. In the first place, it is unintelligible that one would make V and not be able to appraise and affirm value-assessing statements of the kinds "φ manifests v to a greater degree at time t_1 than at t_2," "an institution embodies v to a greater degree at time t_1 than at t_2, or to a greater degree than another institution does," and related ones. This is because, without such assessments, X could not know whether the desire component of X holding a value (item 1 above) was satisfied or not, so that V could not lead coherently to any action (see item 3) but would remain merely a verbal articulation devoid of behavioral significance.

Second, in view of item (2), it is a presupposition of X making V that X endorse that it is possible that φ manifest v highly (or to a higher degree than it does now); for it is unintelligible to affirm the statement "v is a feature of a 'good' φ, but v cannot be highly manifested in φ." *Respecting human rights* is not an sv unless respect for human rights can be highly manifested throughout a society. *The total elimination of injustice* (as distinct from drastically reducing it) is not an sv because it is impossible to totally eradicate injustice. *Predictive power* is not a cv unless there are theories from which predictions can be generated. *Certainty* is not a cognitive value because the scientific methods we deploy cannot generate theories we can know with certainty. Evidence for or against relevant possibility statements may be sought for and obtained in scientific investigations; it will be based in large part on confirmed value-assessing statements. Evidence that counts decisively against the possibility statement constitutes, *ceteris paribus*, a reason to reject V. Thus, the outcomes of scientific inquiry may affect value judgments in logically permissible ways.

Third, in item (2) "good" is functioning as a sort of placeholder. In practice, however, "good" tends to carry with it an ideal of φ — a general, fundamental, comprehensive aim, goal, or rationale for φ of which the φ-values are constitutive (or to which they are subordinate). Such ideals may not be well articulated. In any case, they always remain open to further articulation and, no doubt, to development and revision. When we take them into account (in the next section), the interconnections of value judgments and value-assessing (and other factual) statements become more extensive.

3. Social Ideals and Cognitive Ideals (the Aim of Science)

Consider social values. X might articulate (as I do) as the fundamental ideal or rationale for society that it provide structures that are sufficient to enable all people normally to live in ways that manifest values — that, when woven into an entire life, generate an experience of well-being (fulfillment, flourishing). This ideal, in turn, needs to be complemented by an account of human well-being (a view of human nature) and of how it is intertwined with X's moral ideals. Then v is an sv for X only if X believes that the manifestation of v is constitutive of such structures, or of institutions or movements that aim to bring them about (Lacey 2002b).

Generally, regardless of the social ideal that is articulated, holding v as an sv presupposes that

1. higher manifestation of v contributes to fuller realization of the ideal;
2. higher manifestation of v, and acting to bring it about, do not undermine higher manifestations of other held sv's; and
3. there is not another characteristic that cannot be highly manifested in the same structures as v whose higher manifestation would contribute more toward the realization of the ideal.

These three presuppositions are in addition to the one stated above, which holds that

4. it is possible for v to be more highly manifested (or for its current high manifestation to be maintained) in the relevant society.

And this will be true only if

5. conditions are available (or can be created) in the society that ensure the availability of objects of social value, which are such objects by virtue of their causal contribution to the higher (or maintained) manifestation of v.

All of these presuppositions are open to input from empirical investigation. While they do not entail **V** (the value judgment made when v is held as a value), positive support for them provides backing for one to make **V**, and evidence against them points to the (rational) need to revise either one's sv's or one's articulation of the social ideal. That is significant, even if reaching agreement about the ideal may appear to be out of the question. Suppose, for example, that strong empirical evidence were provided that a valued economic arrangement was a significant causal factor in the perpetuation of widespread hunger, thus showing a causal incompatibility between two plausible sv's. Something would have to give. We are not stuck with across-the-board intractable disagreement. Dialectically, empirical investigation can be crucial in disputes about sv's (Lacey 2002b, 2003b).

The Aim of Scientific Practices

Now turn to cognitive values. A cv is a feature whose manifestation is valued in accepted theories (hypotheses); it is a characteristic of "good" accepted theories. X — qua participant in scientific practices in which theories are entertained, pursued, developed, revised, and appraised — subscribes to an ideal of what makes an accepted theory "good", that is, a general, fundamental, comprehensive aim, goal, or rationale for theories of which cv's are constitutive. While individuals may differ in their judgments about what is constitutive of a "good" accepted theory (that is, about the cv's), so long as there is agreement that theories are intended to be bearers of understanding and knowledge about phenomena, which can (often) be expected to be applied successfully in social practices, the differences will be considered disagreements in need of resolution. Whatever one might maintain about the "subjectivity" of other kinds of values (and my account of values maintains only a little), it is out of place in connection with cv's (Scriven 1974). Since theories are products of scientific practices, I find it convenient to locate the ideal attributed to theories against the background of an ideal of scientific practices.

I will attribute to X the ideal or aim of scientific practices:

A: i. To gain theories that express empirically grounded and well-confirmed *understanding* of phenomena,
ii. of increasingly greater ranges of phenomena, and
iii. where no phenomena of significance in human experience or practical social life — and generally no propositions about phenomena — are (in principle) excluded from the compass of scientific inquiry.

I take understanding to include descriptions that characterize *what* the phenomena (things) are, proposals about *why* that are the way they are, encapsulations of the *possibilities* (including hitherto unactualized ones) that they allow by virtue of their own underlying powers and the interactions into which they may enter, and anticipations of *how* to attempt to actualize these possibilities (Lacey 1999, chap. 5). I have stated the aim so as to encompass all inquiries that are called "sciences" (including social sciences) and those that bear close affinities with them. I include under this rubric all forms of systematic empirical inquiry, because I want neither to rule out by definitional fiat nor to assume a priori that forms of knowledge that are in continuity with traditional forms of knowledge may have comparable cognitive (epistemic) status to those of modern science. Then, I do not restrict what counts as a theory to that which has mathematico-deductive structure or contains representations of laws but also include any reasonably systematic (perhaps richly descriptive or narrative) structure that expresses understanding of some domain of phenomena.[5]

It follows that v is a cv only if it is a characteristic of theories, the sound acceptance of which furthers **A**, and thus only if it is constitutive of a theory's expressing empirically grounded, sound understanding of a range of phenomena. The cv's (that is, gaining theories manifesting them) are constitutive of the cognitive aims of scientific practices. Minimally, then, cv's must in fact — as revealed in interpretive studies that, among other things, explain why they have displaced historically earlier articulations of the cv's (Laudan 1984) — play the role of cognitive aims in the tradition (or traditions) of scientific inquiry and have become manifested in theories whose acceptance is currently uncontroversial.[6] Moreover, if a robust notion of understanding is to be maintained, there should be compelling reasons why other candidates that have been proposed, such as "being an object of social value in the light of some sv," are not included among the cv's.

As already indicated, v is a cv only if it is possible that theories manifest v highly, and hence only if there are available methodological rules that enable its manifestation to be furthered. There are also counterparts of the presuppositions of holding sv's, (a) to (c), that the aim, **A**, renders intelligible. In the present argument, however, I will only draw on the counterpart of presupposition (b) — there are no undesirable side effects of holding v as a cv:

P: a. Gaining higher manifestation of v in a theory is not incompatible with other cv's gaining increased manifestation in that theory; and

b. requiring that v be manifested in an accepted theory does not inhibit entertaining (as potentially worthy of acceptance) theories that might manifest some of the other cv's well.

Nothing said here about conditions on cv's — except perhaps including (b) in **P** — goes beyond Laudan's "reticulated model" (Laudan 1984), which influences everything I have said.

P suffices to eliminate a large range of candidate cv's.[7] The elimination is not achieved a priori, but it is empirically grounded and enacted dialectically. Consider the candidate that it is important for me to eliminate: *being an object of social value in light of specified sv's* (which I will call *OSV*). *OSV* is not a cv. Scientific inquiry in some fields has deployed (more or less explicitly) varieties of it at the cost of inhibiting entertaining theories, which at other times have been shown to manifest well the most commonly touted cv's. These cv's include empirical adequacy, explanatory power, consilience, keeping ad hoc hypotheses to a minimum, the power to encapsulate possibilities of phenomena, and containing interpretive resources that enable the explanation of the success and failures of earlier theories (Lacey 1999, chap. 3). It might be proposed, nevertheless, that there is a particular set of sv's for which this kind of inhibition does (would) not occur. That could only be settled empirically.

Understanding and Utility

Since the outset of modern science, there has been controversy about the relationship of understanding and utility. There has sometimes been a tendency to build utility — that scientific theories have the property of *OSV* for certain sv concerning the control of natural objects — into the aims of science. (Baconian influence has never been absent from the modern scientific tradition.) On most views, however, understanding is a prerequisite for utility; a necessary condition for a theory's having the property OSV is that it manifest the cv's to a high degree, so that from the aim of utility we do not get criteria that can be brought to bear on what constitutes sound understanding. Utility leads us to emphasize certain lines of investigation, requiring that possibilities that may, for example, enhance our capability to control natural objects be addressed and that other lines be neglected because they cannot be expected to give rise to theories that are likely to be objects of social value in light of held sv's. When inquiry is conducted in this way, all theories that come to be accepted will have the property of OSV for these sv's. These theories have OSV because, from the outset of the research process, only theories that would have it, if they were to become soundly accepted, are even provisionally entertained. That a theory actually (and not just potentially) may have the property of utility is established from its being accepted (of a certain domain of phenomena) in view of its manifesting the cv's to a high degree.

The role of utility (OSV) in the selection of theories (when it plays a role) is at the moment where the kinds of theories to be provisionally entertained and pursued are chosen, not where choice among provisionally entertained and developed theories is made (see section 5). What is discovered in the course of investigation is that the theories manifest the cv highly; that they also manifest OSV is simply a consequence of constraints imposed on the inquiry at the outset. If no theories were consolidated under these constraints, we would thereby discover that there are no theories manifesting OSV. OSV does not play a role alongside the cv *at the moment* where theories come to be accepted (section 5). Even if scientists were interested only in investigating useful theories, it would not follow that all theories that can express understanding must manifest OSV. One can build utility into the aims of science (and include OSV among the cv) only by prima facie clashing with **A**. I hasten to add that concluding that OSV is not a cv might be quite trivial if the scientific tradition had not made available an extensive and expanding body of scientific understanding, vindicated by being expressed in theories that have been soundly accepted of certain domains of phenomena, with respect to which they manifest the cv to the highest degree available (Lacey 1999, 62–66). "Utility" has been well served by this knowledge.

Strategies, Methodological Rules, Metaphysics, and Social Values

One might respond that even if, for any sv, OSV is not a cv, nevertheless there may be a set of sv's such that the methodological rule (**MR-OSV**) is an object of high cognitive value.

MR-OSV: Entertain and pursue *only* theories that are constrained so that, if soundly accepted, they also manifest OSV (being an object of social value in light of the specified social value).

If the historical record were to support the proposition that following **MR-OSV** uniquely furthers the realization of the aim of science, **A**, then the distinction between cv and sv's would not be of much importance.

In rebutting this response, I note that **A** pulls us in two directions, which in actual scientific practice can be in tension. One pull is toward putting research effort where we can expect to have readily recognizable success quickly and efficiently. The other is toward engaging in research on phenomena that have been underinvestigated: where, since scientific research builds on its past achievements, theories may appear to be underdeveloped, lacking in internal sophistication, and not very general in scope. Later I will argue that it is important for science, qua worldwide institution, to react to both pulls. I think that the first pull has dominated modern science, however, and that it has indeed led to a methodological rule of the kind just stated being taken as an object of high cognitive value. Furthermore, adopting this methodological rule has enabled some reactions to the other pull, as reflected in the regular emergence of new branches of "modern" science (Lacey 2003b). Nevertheless, I will argue that what is "left out" is of great importance.

Materialist Strategies and the Methodological Rule to Conduct Research under Them

Before stating the methodological rule of the type **MR-OSV** that I have in mind, I will first introduce another one, **MR-M**, which functions in close concert with it but does not, on its face, link the pursuit of science with any social values.

MR-M: Entertain and pursue *only* theories that instantiate general principles of materialist metaphysics about the constitution and mode of operation of the world.

I take materialist metaphysics to affirm that the world (phenomena and the possibilities they permit) can be adequately represented with categories that may be deployed under appropriate varieties of what I call *materialist strategies*, S_M (Lacey 1999, 2002a), so that adopting **MR-M** is tantamount to adopting the methodological rule **MR-S_M**:

MR-S_M: Entertain and pursue *only* theories that meet the *constraints* of S_M, that is, that represent phenomena and encapsulate possibilities in terms that display their lawfulness, and thus usually in terms of their being generated or generable from *law and/or underlying structure, process, and interaction*, dissociating from any place they may have in relation to social arrangements, human lives, and experience, from any link with value, and from whatever social, human, and ecological possibilities that may also be open to them; and (reciprocally) select and seek out empirical data that become expressed using descriptive categories that are generally quantitative, applicable by virtue of measurement, instrumental, and experimental operations — so that they may be brought to bear evidentially on theories entertained and pursued under the constraints.

On this account, materialist metaphysics does not entail physicalism, any form of reductionism, or determinism, and its concrete content is not fixed. Its articulations develop with the developments and changes of S_M — and thus with decisions (which fit Laudan's "reticulated model") about which versions of **MR-S_M** to follow at a given time in a given field — and so is not tied to any particular versions of S_M, certainly not to surpassed mechanistic versions.

Although modern science has been conducted virtually exclusively under varieties of S_M (Lacey 1999), I will suggest that they are just one (albeit an especially important one) among a multiplicity of kinds of strategies that might be adopted in systematic empirical inquiry. That any one of them should be generally prioritized in inquiry does not follow from **A**. Reference to **A** does not provide direction to research, define what counts as worthwhile or significant research, or dictate what questions to pose, what puzzles to resolve, what classes of possibilities to attempt to identify, what kinds of explanations to explore, what categories to deploy both in theories (hypotheses) and observational reports, what phenomena to observe, measure, and experiment on, who the appropriate participants are in research activity and what their qualifications, life backgrounds, and virtues are. None of these matters can be addressed without the adoption of a *strategy* (Lacey 1999, chap. 5, 2002a),[8] whose principal roles are to *constrain* the kinds of theories (hypotheses) that may be entertained in a given domain of inquiry (so as to *enable* investigation) and the categories that they may deploy—and thus to specify the kinds of possibilities that may be explored in the course of the inquiry—and to select the relevant kinds of empirical data to procure and the appropriate descriptive categories to use for making observational reports. Different classes of possibilities may require different strategies for their investigation (see below).

Since, under S_M intentional and value-laden categories are deliberately excluded from use in the formulation of theories, hypotheses, and data, where they are deployed there cannot be any value judgments among the formal entailments of theories and hypotheses. Following MR-S_M thus suffices to ensure cognitive neutrality (see section 1). This is a feature of the design of S_M. There can be no doubt that conducting research under them has been fruitful and that it has furthered **A** to a remarkable extent; there are plenty of theories, expressing understanding of an ever increasing number and variety of phenomena, developed under S_M, that manifest the cv's to a high degree. Moreover, S_M have proved to be highly adaptable, and new varieties of them have evolved as research has unfolded: varieties expressing mechanism, lawfulness expressed mathematically, various forms of mathematical laws (presupposing Newtonian space and time and relativistic space-time, deterministic and probabilistic, with and without physicalistic reductionism, functional and compositional), computer modeling, molecular and atomic structures, among others. Thus, there can be little doubt that research that follows MR-S_M is indefinitely extendable, and we can expect that **A** would continue to be furthered by following it. Minimally, then, MR-S_M' has high cognitive value, where MR-S_M' differs from MR-S_M by dropping the "only" in the formulation.

Bounds of Research Conducted under Materialist Strategies

It does not follow, however, that—even in principle—all phenomena may be understood within the categories that MR-S_M allows. That is because, under S_M, phenomena are investigated in dissociation from any context of value, and so any possibilities that they may derive from such contexts are not addressed. Routine explanations of human action (including those involved in scientific research and applied practices), and attempts to anticipate it, eschew such dissociation. They deploy categories (intentional as well as value ones, including those used in value-assessing statements) that are inadmissible under S_M, and theories containing them may manifest the cv's highly (Lacey and Schwartz 1996; Lacey 2003b).

Currently, although there is no compelling evidence that these theories are reducible to or replaceable by theories constructed under S_M, varieties of S_M have been adopted in the neuro-, behavioral, cognitive, and other sciences anticipating that reductions or replacements will be found. This illustrates science, following MR-S_M, responding to the pull (the second pull of **A**) to grasp new kinds of phenomena. While theories in these fields have been soundly accepted of some (mainly experimental) domains of phenomena, that they may be extended into characteristic phenomena of human agency remains simply an anticipation with slender empirical warrant (but often fostered by metaphysical materialist theories of the mind). The conjecture that adequate understanding of human agency can be produced under S_M is worth exploring, and I do not a priori rule out its eventual confirmation, but it can only be tested rigorously against the products of research conducted under strategies (call them "intentional strategies," S_I) that do not make the dissociations made by S_M. To limit research to that conducted under S_M would inhibit the exploration of theories under which, for example, *explanatory adequacy* might become highly manifested. Then, furthering of **A**—such as in connection with the investigation of value phenomena—would require deployment of strategies (say, S_I) under which such theories could be entertained and appraised (Lacey 1999, chap. 9; 2003b).

Why, then, has research conducted under S_M been so dominant—so much so that these strategies have often been seen as constitutive of scientific research so that research that does not adopt them (as in some social sciences) is considered not quite "scientific"? Appeal to materialist metaphysics does not justify it (but see, for comparison, Bunge 1981), for it lacks the necessary cognitive (empirical) credentials. Could it be that the first pull of **A** is so strong that, so long as S_M continue to expand into new fields, scientists are content to limit themselves to gaining the understanding they are confident of gaining now, and to leave it to the future (anticipating that still more sophisticated versions of S_M will become available) to address phenomena that are unexplored under these strategies now? Clearly, this suggestion, though important for explaining the dominance of research under SM, cannot be sustained as a justification when one considers the application of scientific theories.

The Importance of Considering Applications

I have already indicated that theories may be appraised not only for their cognitive value but also for their social value. Can they, on application, inform projects valued in view of specified sv's? The traditional answer is yes, because any theory that is applicable at all can inform projects of any viable sv — theories are neutral (see section 1).[9] But I contend that this cannot be sustained. As pointed out earlier, theories developed under S_M are indeed cognitively neutral. But they need not thereby be neutral-in-application. An example will help to illustrate this point. Consider transgenic seeds, which are embodiments of soundly accepted theoretical knowledge developed under versions (biotechnological) of S_M. As technological objects, however, they have no significant role in the projects of those who aim to cultivate productive, sustainable agroecosystems in which both biodiversity is protected and local community empowerment is furthered. Hence they have little social value for the many rural grassroots movements throughout Latin America (and elsewhere) that hold these values. Their applications do not display evenhandedness. Nevertheless, projects aiming to further these values do not lack scientific input. Theories that inform them are consolidated by research conducted under *agroecological strategies* (S_{AE}), strategies under which a multiplicity of variables (concerning crop production, ecological soundness, biodiversity, and local community well-being and agency) are investigated simultaneously and interactively.[10]

What is included in the range of sv's, for which a theory may have social value, is a matter of fact, open to empirical inquiry. I believe that there is a measure of neutrality-in-application among theories confirmed under S_M taken as a whole, yet when S_M are pursued virtually exclusively, they have social value generally and especially in light of value-outlooks that highlight a set of social values that concern specifically modern ways of valuing the control of natural objects (*mcv*). These values concern the scope of control, its centrality in daily life, and that relatively *mcv* are not subordinated to other moral and social values — so that, for example, the expansion of technologies into more and more spheres of life and as the means for solving more and more problems is especially highly valued, and the kind of ecological and social disruption this causes is taken to be simply the price of progress.[11] Theories confirmed under S_M tend to be especially pertinent to inform projects valued in light of *mcv* (and, in many cases, like the transgenics one, only them).

Materialist Strategies and Their Links with Specifically Modern Ways of Valuing Control

This fact, the special pertinence of theories confirmed under S_M in light of *mcv*, helps *explain* the predominance of research conducted under S_M and the kind of epistemic privilege attributed to theories soundly accepted under them, for *mcv* are widely held in modern societies and reinforced by their relationships with other values that are highly manifested in powerful contemporary social institutions (generally at the present time linked with capital, the market, and the military). The explanation I propose, however, is not the simple one that S_M are adopted for the sake of generating applications that further interests cultivated by *mcv*. (In particular inquiries they are often adopted for quite different reasons.) Rather it is because there are several ways in which holding *mcv* and adopting S_M *mutually reinforce* each other (Lacey 1999, chap. 6; 2002a). Here is what they involve: (1) The furtherance of *mcv* is served by and dependent on the expansion of understanding gained under S_M; (2) there are close affinities between technological and experimental control (for comparison, see Dupré 2001, 10–11), and understanding, gained under S_M, is dependent on successfully achieving experimental control; (3) engaging in research under S_M fosters an interest in the fuller manifestation of *mcv*, since its pursuit often depends on the availability of instruments that are products of technological advances linked with *mcv*, and sometimes technological objects themselves provide models or even become central objects for theoretical inquiry; and (4) given current forms of the institutionalization of science, where the institutions that provide material and financial conditions for research are likely to do so because they expect that applications "useful" for them will be forthcoming, any values furthered by research under S_M (such as the satisfaction of curiosity and other values associated with "basic" research) tend to be manifested today within value-outlooks that also include *mcv*.

It is through these mutually reinforcing relations that theories soundly accepted under S_M become objects of social value in light of *mcv*. (Developing my earlier notation, these theories manifest *OMCV*; and with respect to other *sv*'s — for example, those of the rural movements referred to earlier — they may not manifest *OSV* significantly.) It is a condition of their manifesting *OMCV* highly and a ground for expecting the *efficacy* of their applications, of course, that their cognitive value is confirmed under S_M.

That S_M have become the predominantly used research strategies rests on the fact that theories, which become soundly accepted under S_M, also manifest *OMCV*,[12] and thus effectively are products of following the methodological rule (an instance of **MR-OSV**):

MR-OMCV: Entertain and pursue *only* (virtually exclusively) theories that are constrained so that, if soundly accepted, they also manifest *OMCV*.

MR-OMCV is the methodological rule (which I referred to at the outset of this section) that is widely taken to be an object of high cognitive value among the practitioners of modern science.

The link between S_M and *mcv* not only explains but also provides the bulk of the rational grounds (if there are any) for the predominance given to research under S_M. Thus, I suggest that not only is adopting **MR-M** tantamount to adopting **MR-S_M** but also that adopting **MR-S_M** is tantamount to adopting **MR-OMCV**. According primacy to research conducted under S_M is not derived simply from commitment to furthering **A**. Research conducted in this way has enabled **A** to be furthered to a remarkable degree, while its products have informed the processes that have so entrenched the manifestation of *mcv* in the leading contemporary socioeconomic institutions. Rationalizing the predominance of research conducted under S_M depends on the rational endorsement of *mcv* (perhaps reinforced by commitment to other *sv*'s, such as those of the market; see Lacey 2002d).

The Need for a Multiplicity of Strategies

Many people, perhaps awed by the increasing manifestation of *mcv*, which seems to be integral to the trajectory of current predominant economic forces, unreflectively assume that *mcv* contain a set of universal values, so that they recognize no difference in applicability between being in service to *mcv* and to all viable value-outlooks. Both *mcv* and materialist metaphysics are deep in the unreflective consciousness of educated people in advanced industrial nations and their allies, so much so that few question these notions. Thus, it seems apparent to them that scientific research is identical to research conducted under S_M, and furthering **A** reduces to carrying out research under S_M.

In contrast, sometimes there is a choice of strategies to be made, such as in agriculture, as I pointed out earlier, for S_M lack the resources needed to explore certain classes of possibilities. In such situations, furtherance of **A** would require that research be done, within the collective body of scientific institutions, under both S_M and competing strategies for whose potential fruitfulness there is some empirical support. When competing strategies are not pursued, I have just suggested, the key factor is endorsement of *mcv*. Of course, for those who reject *mcv*, this does not count as legitimation, and it poses no rational barrier for them to adopt different strategies that promise to provide knowledge and to identify possibilities that may inform projects with roots in their own values (such as adopting S_{AE} in view of their links with the values of rural organizations).[13]

Efficacy and Legitimation

It is clear enough that under S_M possibilities are identified such that, when they are realized in technological objects or other interventions, the interests of *mcv* and the *sv*'s linked with them (not always only them) are likely to be enhanced. Generally, application presupposes efficacy: that a certain intervention or technological object with a specified design will work effectively, that it will perform as intended. Theories soundly accepted under S_M can be counted on to be an abundant font of efficacious applications of value in light of *mcv*, whereas theories developed under competing strategies, whatever efficacy they might engender, tend to further competing social values. The role of *mcv* is even more far-reaching, however, as will become clear when we take into account that application involves accepting hypotheses not only about efficacy but also about matters that underlie legitimation.

Legitimation not only involves consideration of the social value that will be directly furthered by an efficacious application but also requires the backing of the hypotheses of the following kinds:

No serious negative side effects (NSE): There are no side effects — of significant magnitude, probability of occurrence, unmanageability — of negative social value caused by the application.

No "better" way (NBW): There is no other way, potentially of greater social value, to achieve the immediate goals of the application (or competing goals with greater social value).[14]

In practice, the importance of *NSE* is generally recognized and no application is introduced without some attention to it. Three interconnected things should be noted. First, addressing *NSE* is usually assigned to studies of risk analysis. Risk analysis involves making conjectures (based on theoretical considerations or suggested by observations) about possible negatively valued side effects and then designing studies — in which specific problems, normally open to investigation under S_M, are posed — to determine their likelihood, manageability, and so on. These are problems where answers are not settled simply by assessments of cognitive value given available empirical data; matters of social value are also essentially involved (see discussion in section 5, "endorsing" hypotheses). Moreover, questions about the generality of proposed answers and of unknown risks are always hovering in the background, and this kind of risk-assessment has no means to explore the effects of applications, qua objects of social value under specific socioeconomic conditions. Second, *NSE* is of negative existential logical form. Evidence for it is largely the failure to identify empirically well-supported counterinstances. (It may also gain evidential support from theories in which it may be embedded.) But *absence* of identified counterinstances is not the same as *failing to identify* them. Relevant failure is failure after having conducted appropriate investigation, and "appropriate" is a value-laden term. Since *NSE* cannot be formulated using only the categories of S_M, the absence of systematic empirical inquiry that pursues theories in which it is contained may leave the attempt to identify counterinstances a pretty haphazard affair. Third, where *mcv* are held, and research conducted under S_M privileged, the "burden of proof" is placed squarely on critics once standard risk assessment has been carried out. Yet, often, in view of the way in which science is institutionalized, the conditions are not available for them to assume the burden. The consequence is that *NSE* may go unchallenged, even though, were the conditions available, a cognitively well-based critique might be able to be developed. *NSE* does get on the agenda, however, and standard risk assessments are the response, and in numerous cases that is quite adequate.

Where S_M are predominant, *NBW* does not even make the agenda in connection with some technological innovations. Where a proposed alternative way is linked with *sv*'s that are in tension with *mcv*, the tension tends to be considered sufficient for rejecting these *sv*'s and dismissing the alternative out of hand. When this happens, no serious consideration is given to whether conducting research under strategies linked with these *sv*'s might generate knowledge that would fruitfully open up new possibilities that (unlike those identified under S_M) could serve to further interests fostered by them.[15] This is inconsistent with taking **A** to express the aim of science. **A** does not rest easily with limits on the scope of neutrality-in-application — even when that neutrality is subordinated only to a value of such widespread acceptance as *mcv*. Neither neutrality-in-application nor adequate appraisal of *NSE* and *NBW* can be achieved without research being conducted under a variety of strategies, respectively linked with different *sv*'s.

Impartiality of Theory Acceptance but Not Autonomy of Methodology

A methodological rule has cognitive value if it contributes to furthering **A**. As already pointed out, **MR-S_M'** clearly has cognitive value in connection with exploring many phenomena and possibilities. **MR-S_M**, insofar as it encompasses **MR-S_M'**, contributes to furthering **A** (i–ii); but insofar as it goes beyond it by virtue of the "only," it tends to undermine **A** (iii). However, adopting S_{AE} furthers (iii) — and also (i) and (ii), even if quantitatively to a much smaller extent than adopting S_M.[16] There are certain classes of possibilities that, so far as we know at present and can expect in the foreseeable future, cannot be identified by research conducted under S_M — for example, potential objects of social value given specified *sv*'s (such as those of the rural movements), and those that might be identified under another strategy (S_{AE}). It is holding these *sv*'s that makes it especially interesting to adopt these strategies, just as (if I am right) it is holding *mcv* that grounds adopting only S_M. Conflict about *sv*'s thus impinges on what methodological rules to follow, and thus *autonomy* of methodology (section 1) cannot be upheld. But *impartiality*, as an approachable ideal of accepted theories, is left untouched. Many theories, developed by following **MR-S_M**, have been soundly accepted of wide-ranging classes of phenomena by virtue of their manifesting the *cv*'s to a high degree. This is a fact that is unaffected by the virtual equivalence of **MR-S_M** and **MR-OMCV**. A similar logic applies where other strategies are chosen.

5. Moments of Scientific Activity

Out of this discussion a model of scientific activity emerges in which it is useful to distinguish (analytically not temporally) three moments of scientific activity: M_1, adopting a strategy; M_2, accepting theories; and M_3, applying scientific knowledge.

To accept a theory T is to deem that T needs no further testing and that T may be taken as a given in on-going research and social practice. According to impartiality (now supplementing, with positive content, the negative characterization given in section 1), T is soundly accepted of a specified domain of phenomena if and only if it manifests the cv's to a high degree and if, given current "standards" for "measuring" the degree of manifestation of cv's, there is no plausible prospect of gaining a higher degree (Lacey 1999, chap. 1; 2002a). Given **A** as the aim of science, and given that it is in theories that understanding of phenomena is expressed, there is no rationally salient role for sv's at M_2; the fact that T may manifest some OSV highly counts rationally neither for nor against its sound acceptance.

At M_1 and M_3, however, sv's have legitimate and often rationally indispensable roles. At M_3 obviously an application is made because it is intended to serve specific interests, and thus to further the manifestation of specific sv's, and judgments of its legitimation depend on a multiplicity of value judgments. At M_1 a strategy may be adopted (as we have seen) — subject, in the long run, to research conducted under it being fruitful in generating theories that become soundly accepted at M_2 — in view of mutually reinforcing relations between adopting it and holding certain sv's, as well as the interest in furthering those values. (Sometimes it may be adopted for other reasons.) Adopting a strategy defines the kinds of possibilities that may be identified in research — in important cases possibilities that, if identified and actualized, would serve interests cultivated under the sv's linked with adopting the strategy. Adopting a strategy per se does not imply that possibilities of these kinds exist, nor, if they do exist, concretely what they are; such matters can only be settled at M_2 where impartiality should be adhered to.

Thus, neutrality-in-application cannot generally be counted on to hold; on application, at M_3, theories tend to serve especially well the sv's linked with the strategy under which they are accepted. Nevertheless, neutrality-in-application should remain an aspiration in the institutions of science, but now it should be given the following interpretation: **A** should be pursued in such a way that scientific knowledge is produced so that projects valued in light of any viable sv's can be informed, more or less evenhandedly, by well-established scientific knowledge. Generally, that neutrality-in-application is lacking in actual fact will be testimony to the fact that **A** has been pursued largely in response to the first pull that I identified. The thoroughgoing pursuit of **A** requires adopting multiple strategies. I doubt that this can happen without the role of sv's at M_1 being recognized as legitimate, and strategies linked with the sv's of less dominant groups being provided with appropriate material and social conditions for their development (Lacey 2002c,d).

Roles for Values at the Different Moments

Before concluding, I elaborate some points pertinent to the role of sv's at the different moments. In the first place, although OSV, regardless of the sv's that may be considered, does not play a proper logical role alongside the cv's at M_2, sv's may play various roles at this moment. Here are examples: (1) Institutions that manifest certain sv's may have cognitive value; (contingently) they may provide conditions necessary for accepting theories in accordance with impartiality. (2) That sv's are held may partially explain why accepted theories of certain domains of phenomena are available but not of others. (3) Adequate testing of theories — and especially the specification of the limits of the domains of phenomena of which they are soundly accepted — may require critical comparison with theories developed under a competing strategy that has mutually reinforcing relations with particular sv's. (4) Since a theory may not be neutral-in-application or may undercut the presuppositions of a value outlook, commitment to sv's (that are not served by the theory's application or whose presuppositions are undercut) can lead to raising the "standards" for "measuring" the degree of manifestation of the cv's. (5) Holding particular sv's may attune us to diagnosing when a theory is being accepted in discord with impartiality, such as when an OSV is in fact covertly playing a role at M_2 alongside the cv's.[17]

Furthermore, accepting/rejecting is not the only stance relevantly taken toward T in scientific activity. T may be provisionally entertained, pursued with a view toward its development or revision, subjected to testing, held to be more promising or to "save the phenomena" better than extant alternatives, used instrumentally in other inquiries, and so on. Clearly some of these stances must have been adopted at earlier stages of the research processes that produce a soundly accepted theory. (Some theories are never candidates for acceptance — "ideal" theories, some mathematical "models.") The model of scientific activity proposed here may be elaborated to include further moments and submoments to correspond to these stances. At some of these moments, sv's may have proper roles. Once a strategy has been adopted at M_1, for example, there is a moment at which specific problems for investigation are chosen. Even those who endorse autonomy (and who do not recognize that there is a question of choice of strategy) readily admit a role for sv's at this moment.

Finally, I have maintained that application is an important moment (M_3) of scientific activity, so much so that the sv's served by application may also play a role at the moment (M_1) at which a strategy is adopted. Thus, sv's play a role at the heart of scientific activity, and I see no good reason to eliminate them from this role. At M_3, sv's also play a variety of roles connected with the legitimation of applications. Legitimation requires attention to NSE and NBW, and scientists — as scientists — are expected to make judgments about them (Lacey 2003b). These judgments are not reducible to those of theory acceptance. I will say that scientists *endorse* (or do not endorse) hypotheses of the types NSE and NBW. To endorse a theory or hypothesis involves appealing to both cv's and sv's.[18] Those who want to keep sv's out of the heart of scientific activity do not consider judgments of endorsement to be proper scientific judgments. This is a pretty implausible claim (Machamer and Douglas 1999; Douglas 2000). Scientists do and are expected to tackle problems connected with NSE and NBW.

To endorse T is to judge that T has sufficient cognitive value (that is, it is sufficiently likely to be true, or to become soundly accepted) that the possibility of its being false or of future research leading to its rejection, and the possible consequences (serious negative ones from the perspective of specified sv's) of acting on it if this were to happen, should not be considered good reasons for not engaging in actions informed by T. Endorsing T is a necessary (but not sufficient) condition for the legitimation of its application. Acceptance implies endorsement (Lacey 1999, 71–74), but we cannot reasonably expect acceptance as a general condition on endorsement: we *have* to act in the absence not only of certainty but also of knowledge meeting the high standards needed for sound acceptance. This is particularly relevant where matters like *NSE* and *NBW* are pertinent. When we endorse without acceptance, *sv*'s are always in play, whether recognized consciously or not, and scientists' judgments may differ because of the different *sv*'s they hold (Douglas 2000). Endorsement is an important moment of scientific activity. To deny endorsement to *T*, when other scientists endorse it, carries the obligation to specify what further testing is needed (Lacey 2002b,c,d). If none can be specified after a due lapse of time, then *T* has been tested according to the highest and most rigorous available standards. Science can do no more.

6. Conclusion

I am indeed interested in theoretical products of science having sound cognitive credentials. I share this interest with the traditional view. But I do not want to keep (noncognitive) values out of science. Values are already there. What I am recommending is reform. I want to let in more values *in the proper place* and thus to dull the influence of the values that are already there. No doubt, my recommendations may allow some unsavory characters (posing as my friends) to get a foot in the door, but the model of scientific activity I propose puts plenty of resources behind slamming the door hard on them if need be. This model enables us to understand certain phenomena of current scientific activity and to underwrite recommendations about how to improve it. It requires distinguishing between *cv*'s and *sv*'s and reflects cognitive as well as political interests.

NOTES

I am grateful to Peter Machamer for his commentary at the conference. It stimulated me to clarify several parts of this chapter.

1. The view *science is value free* sketched here and its three constituents (impartiality, neutrality, and autonomy) is elaborated in Lacey (1999, 2002a). See also section 5.

2. For a more complete account of my views about values, see Lacey (1999, chap. 2) and Lacey and Schwartz (1996).

3. In Lacey (1999, chap. 3) I introduce a more general notion of "cognitive value" that applies to beliefs held in ordinary life and inform actions in a variety of practices. The narrower notion is sufficient for the purposes of this chapter; as I use it here, "cognitive value" may be considered an abbreviation of "value of accepted theory."

4. McMullin (1983) and Nagel (1961), using different terminologies, deploy a similar distinction.

5. Stating the aim as I have is, of course, contentious. Laudan (1977), for example, states it in terms of "problem solving." I simply note that much of my argument can be rearticulated in the context of other accounts of the aims of science.

6. Another question may be raised here. Let me put it in a quasi-paradoxical form: Even if cv's are distinct from other kinds of values, is it appropriate to deploy the same set of cv's regardless of the domain of inquiry (for example, physical or social) or the strategy (section 4) adopted in research? Or could it be argued that the cv's are in some sense relative to the strategy adopted? This question needs further exploration, and the answer to it could affect my current views. Peter Machamer, commenting on a draft of this chapter, raised what I think is a related issue. He wrote, "The relevant reliability of knowledge lies in the certification by public social norms." I take this to imply that a discourse that aims to identify cognitive values already expresses certain social values, so that — apparently paradoxically — any nontrivial distinction drawn between cognitive and social values will in fact be "relative to" these particular social values. Both of these matters are important, but addressing them is outside the scope of this chapter.

7. The range includes commonly identified "outside interferences" that have been considered to threaten the autonomy of the conduct of science: *accepted by consensus, popularly believed, consistency with the presuppositions of particular social values, consistency with biblical interpretations or with the tenets of dialectical materialism*. Consistency with biblical interpretations, for example, clashes with *empirical adequacy, keeping ad hoc hypotheses to a minimum,* and *power to encapsulate the possibilities permitted by actual phenomena.*

8. What I call "strategies" has much in common with Laudan's "research traditions" (Laudan 1977), Kitcher's "frameworks" (Kitcher 1993, 57), and Hacking's "form of knowledge" (Hacking 1999, 170–71).

9. In different ways Kitcher and Longino question *neutrality*. Kitcher does so by raising questions about how science bears on human flourishing (Kitcher 1993, 391; 1998, 46), and he has developed this with his recent reflections on the cognitive and social aspects of the *significance* of scientific results (Kitcher 2001, chap. 8); Longino does so partly by questioning whether there is a significant distinction between cognitive and social values (1990, 2002).

10. The claims simply asserted in this paragraph are all elaborated and documented in detail in a series of recent analyses (Lacey 2000, 2001, 2002c,d, 2003a; also 1999, chap. 8). My account of S_{AE} draws heavily on Altieri (1995). Their role here is illustrative; the thrust of my general argument does not depend on agreeing with what I claim about them. Illustrations could be drawn just as readily from considerations involving S_I (Lacey 2003b) and research in the social sciences (Lacey 2002b).

11. See Lacey (1999, chap. 6; 2002a) for details behind this summary statement. In these works I list explicitly the values that jointly constitute *mcv*, explore the presuppositions of their legitimation, and provide evidence for all the assertions that follow in the next subsection.

12. See Lacey (1999, chaps. 5, 6; 2002a) for the detailed argument for this explanation and against other proposed explanations.

13. For the adherents of *mcv*, the values of contesting movements are usually held not to be viable, since they are considered to represent either abhorrent social visions or unattainable ideals—and the contestants trade comparable charges (Lacey 2002b,c).

14. All the details of this subsection have been illustrated in detail in my recent works on the interplay of ethics and the philosophy of science in current controversies about transgenics and on the promise of agroecology. See the references in note 10. Regarding efficacy, research and applied practice show that some transgenic crops are productive (have equivalent or greater yields than conventional alternatives), resistant to specified herbicides and pests, and so on. And proponents of transgenics maintain and the critics deny (regarding *NSE*) that plantings and consumption of transgenic crops and products will not directly cause harm to human health and the environment; and (regarding *NBW*) that there are no other (nontransgenic-intensive) forms of agriculture that can be sufficiently productive to meet the world food needs in coming decades.

15. It happens in the controversies about transgenics, in which the possibilities of agroecology are often dismissed on exactly this ground; and this, of course, runs counter to impartiality. (See the references in note 10 for documentation.) See also note 13. On the fruitfulness of S_{AE}, see especially Altieri (1995) and references in Lacey (2002c). One of the greatest risks posed by the widespread implementation of transgenic-intensive agriculture is that it will undermine the conditions needed for both research and practice of agroecology. Acting vigorously, claiming legitimacy on the ground of *NSE*, may *bring it about* that *NBW becomes* true.

16. I believe that SM have special importance in scientific practices (Lacey 1999, chap. 10); but, for the sake of furthering **A**, they should be complemented by other strategies. Strategies like S_{AE} draw on the results of research conducted under S_M in numerous indispensable ways. They should not be seen as complete alternatives to S_M, but more as an interlocking set of local approaches each of which draws on the results of S_M where convenient.

17. Longino (1990; 2002) deploys items similar to these when arguing against a significant distinction between the cognitive and the social. I cannot engage her arguments here, but see Lacey (1999, chap. 9).

18. The category "value appealed to in offering solutions to problems addressed by scientists" does not provide a ground for the distinction between *cv*'s and *sv*'s. Even so, the distinction is important in the context of appraising *NSE* and *NBW*. It underlies how disagreements about endorsement can be rationally addressed and helps to explain why often they are not (see references in note 10).

REFERENCES

Altieri, M. 1995. *Agroecology: The science of sustainable development*. 2nd edition. Boulder, CO: Westview Press.

Bunge, M. 1981. *Scientific materialism*. Dordrecht: Reidel.
Douglas, H. 2000. "Inductive risk and values in science." *Philosophy of Science* 67:559–79.
Dupré, J. 2001. *Human nature and the limits of science*. Oxford: Clarendon Press.
Hacking, I. 1999. *The social construction of what?* Cambridge, MA: Harvard University Press.
Kitcher, P. 1993. *The advancement of science: Science without legend, objectivity without illusions*. New York: Oxford University Press.
———. 1998. A plea for science studies. In *A House built on sand: Exposing postmodernist myths about science*, ed. N. Koertge, 32–56. New York: Oxford University Press.
———. 2001. *Science, truth, and democracy*. New York: Oxford University Press.
Lacey, H. 1999. *Is science value free? Values and scientific understanding*. London: Routledge.
———. 2000. Seeds and the knowledge they embody. *Peace Review* 12:563–569.
———. 2001. Incommensurability and "multicultural science." In *Incommensurability and related matters*, ed. P. Hoyningen-Huene and H. Sankey, 225–39. Dordrecht: Kluwer.
———. 2002a. The ways in which the sciences are and are not value free. In, *In the scope of logic, methodology, and philosophy of science*, vol. 2, ed. P. Gardenfors, K. Kijania-Placek, and J. Wolenski, 513–26. Dordrecht: Kluwer.
———. 2002b. Explanatory critique and emancipatory movements. *Journal of Critical Realism* 1:7–31.
———. 2002c. Assessing the value of transgenic crops. *Ethics in Science and Technology* 8:497–511.
———. 2002d. Tecnociência e os valores do Forum Social Mundial. In *O espírito do Porto Alegre*, eds. I. M. Loureiro, M. E. Cevasco, and J. Corrêa Leite, 123–47. São Paulo: Paz e Terra.
———. 2003a. Seeds and their socio-cultural nexus. In *Philosophical explorations of science, technology and diversity*, ed. S. Harding and R. Figueroa, 91–105. New York: Routledge.
———. 2003b. The behavioral scientist *qua* scientist makes value judgments. *Behavior and Philosophy* 31.
Lacey, H., and Schwartz, B. 1996. The formation and transformation of values. In *The philosophy of psychology*, ed. W. O'Donohue and R. Kitchener, 319–38. London: Sage.
Laudan, L. 1977. *Progress and its problems: Toward a theory of scientific growth*. Berkeley: University of California Press.
———. 1984. *Science and values: The aims of science and their role in scientific debate*. Berkeley: University of California Press.
Longino, H. E. 1990. *Science as social knowledge*. Princeton, NJ: Princeton University Press.
———. 2002. *The fate of knowledge*. Princeton, NJ: Princeton University Press.
Machamer, P., and Douglas, H. 1999. Cognitive and social values. *Science and Education* 8:45–54.

McMullin, E. 1983. "Values in science." In *PSA 1982*, vol. 2, ed. P. D. Asquith and T. Nickles, 3–28. East Lansing, MI: Philosophy of Science Association.

Nagel, E. 1961. *The structure of science.* New York: Harcourt, Brace, and World.

Scriven, M. 1974. The exact role of value judgments in science. In *Proceedings of the 1972 Biennial Meeting of the Philosophy of Science Association,* ed. R. S. Cohen and K. Schaffner, 219–47. Dordrecht: Reidel.

3

Epistemic and Nonepistemic Values in Science

Mauro Dorato
Department of Philosophy, University of Rome

In a preliminary investigation of the complex and uneven territory covered by the relationship between science and values, it is important to distinguish different ways in which values can enter into the natural and social sciences. Very often these differences have been conflated, to the detriment of our understanding of the geography of the territory.

I contend that science's value dependence has at least *four* different forms, which give us four different roles that values play in the scientific endeavor: (1) values functioning as selectors among different fields of investigation; (2) values functioning as selectors among alternative, empirically equivalent theories or hypotheses (these are often referred to as *epistemic values*); (3) values functioning as "regulative" ideas of science, that is, as indicators of the place and meaning that the scientific enterprise as a whole should have in our culture, in society, and in our life in general; (4) values functioning as guides to the application of our scientific knowledge and technology to practical decision making. Under the fourth role I classify cases in which the piece of knowledge to be applied is highly reliable as well as cases in which our knowledge is instead partial and limited, generating nonnegligible "inductive risks" in the sense introduced by Carl Hempel (1965) and further developed by Heather Douglas (2000).

By reviewing what I regard as the main problems in each of these different types of science's value dependence, I offer a *general survey of the territory while stressing the importance of keeping epistemic and nonepistemic values separate*. The cases in which they cannot be easily separated are those in which our knowledge is quite unreliable. My use of the word *importance* implicates me as an advocate of separating epistemic values from nonepistemic ones, a position I intend to defend against current attacks. By *epistemic values* I am referring not just to the evidence-hypothesis relationship or to values that are conducive to *truth*[1] but more generally to those aims that are usually regarded as capable of furthering our knowledge, like "understanding" or "explaining," to the extent that these two are independent of each other. I use *nonepistemic* to refer essentially to values that are ideological, economical, or political (like feminism, sexism, Marxism, fascism, capitalism, and racism), or ethical, environmental, esthetical, or religious.

Sometimes such nonepistemic values are grouped under the misleading term *social values*. Here, however, I will simply take for granted that all the typically epistemic values I will discuss (consistency, experimental accuracy, explanatory power, and so on) are *socially shared*, because they are an essential part of the training of scientists.[2] By making this assumption, it seems to me that bringing into the discussion the adjective *social* and contrasting it with *epistemic* can only cause confusion and misunderstandings. The interesting contrast is not between the epistemic and the social but rather between the epistemic and the ideological, as already clarified. Inquiring into the contrast between the epistemic and the ideological entails raising the question whether scientists' acceptance of a hypothesis (its warrant) depends on their allegiance to values furthering our knowledge or instead to values pertaining to their political or ideological agendas. Whether these values can and ought to be separated is the object of my investigation here.

I will say little on the first type of dependence since, as we shall see, it is highly noncontroversial and universally recognized: political, military, economic, and, less grandiosely, purely personal and idiosyncratic interests or values clearly influence the choice of scientific problems and facts to be investigated. Though in this sense science is obviously value-laden, such a form of value dependence is not equivalent to claiming that the *cognitive content* of science is laden with nonepistemic values (section 1). As to the second type of dependence, I will argue that nonepistemic values typically do not have (a descriptive claim), and ought not to have (a prescriptive one), any role in the choice among empirically equivalent theories. Nonepistemic values can at most motivate the *pursuit* of a theory, but they *should* never be regarded as a *justification* for choosing one among a class of empirically equivalent theories. The "should" of the previous sentence clearly shows that the claim it expresses is a *normative* one; however, I would also like to contend that the norm in question is *de facto* accepted by most good scientists.

In addition, a widespread misunderstanding of the relationship between theory and evidence has greatly exaggerated the threat posed to the epistemic claims of science by the alleged existence, in any historical situation, of mutually incompatible but empirically equivalent theories. Curiously enough, scientists often hold that, rather than having to constantly and arbitrarily choose between different but empirically equivalent theories, as some philosophers would have them do, they do not even have *one* theory compatible with all known data! Given the clash of these views, something is clearly amiss here (section 2).

The third type of dependence would seem to be the ideal home for the influence of nonepistemic values on science; for instance, whether science ought to aim at truth and true explanations or merely at empirical adequacy or whether its knowledge claims exhaust all we can know about the world seem to be questions imbued with nonepistemic values. On closer look, however, we see that the *evidential strength* or plausibility of, say, the various realist or antirealist scientific and/or philosophical *arguments* do not depend on one's allegiance to such values, so the arguments themselves can and ought to be regarded as free from nonepistemic values. The general argument I will rely on in this context is that the causal origin of *any* argument — the reason why anybody would want it to be valid — has nothing to do with its validity (section 3).

As to the last type of dependence, I will discuss Douglas's (2000) recent analysis, according to which the nonepistemic values involved in inductive risks do not merely belong to the problem of applying our (more or less uncertain) knowledge to action but concern as well the epistemic import of the hypothesis at stake (section 4). Finally, I will also show how these four different ways science depends on values, despite their logical independence, intersect more than once.

In a word, my main thesis is that cognitive values are the only values that *ought* to belong to the *justification* of a scientific theory, or, differently put, to the relationship between evidence and theory. In all cases in which scientists are nevertheless forced by insufficiently reliable knowledge to rely on nonepistemic values to formulate empirical hypotheses, I will show that the resulting claims remain wholly objective (that is, intersubjectively valid).

Since my main claim — that science *is*, in its best examples, free from nonepistemic values and *ought* to remain this way — reflects in part an evaluative attitude toward science, those who disagree with me partially disagree on what the general aims of science should be. Consequently, the dispute about the independence of science's cognitive claims on noncognitive values belongs to the *third* section of this chapter (the third type of dependence in my classification). Being a *normative* thesis about the general aims of science, such an independence thesis cannot be attacked (or defended) *solely* by invoking *empirical* data coming from the history or the sociology of science.[3] Of course, I could be accused of defending too idealized an image of science, and I hope to respond to this criticism in the last section of the chapter.[4]

1. Values as Selectors of Problems: The Choice of Facts

Granting the obvious point that values in general (what we care about) have, among other things, a *selective function* — that is, they help us to choose among different alternatives or possible states of affairs — in the *first* of the four senses mentioned above, individual or social values of all sorts help us to select what we should study, what is worth our scientific investigations, or what, in a broad sense of the words, *looks interesting and important for us*. We are not merely seeking truth *simpliciter*, or even the *whole* truth; what we are after is *significant, interesting truths*: in most cases, we do not care about finding the number of grains of sand on a given time of the day on the shores of planet Earth, even though there certainly is fact about it.

In this first type of science's value dependence, a particular value, that is, a particular *interest* in a certain area of science rather than in another, can "drive" a particular scientist (or groups of scientists) to select that area as his or her own field of expertise. For instance, a young researcher may think it much more promising to devote herself to solid-state physics rather than to particle physics, because there may be much more money in the former field than in the latter (economic, nonepistemic interest), and because teachers operating in solid-state physics may be much more prestigious and fun to work with than those operating in particle physics. In my schematic distinction, all these would count as nonepistemic interests. The young researcher's career may therefore be better fostered in one field than in the other, and her choice might reflect this.

However, within this kind of dependence of science on nonepistemic values, psychological preferences like ambition, desire for social recognition, and expectation of gain typically enter science from "outside" or in an *external* way, namely *by simply stimulating or encouraging the pursuit and development of certain areas of inquiry rather than others*.

Interestingly, the role of nonepistemic values as selectors of interest-relative facts has been first realized by the methodologists of the social sciences. It is this type of value-ladenness of science that Max Weber referred to when he stressed the role cultural values play in making us adopt a particular "perspective" from which to inquire into a certain area of research (say, the Reformation). He called this phenomenon *Wertbeziehung* ("reference to values"): we are free to adopt an *economic, moral,* or *religious* point of view or any other relevant perspective in order to illuminate the historical phenomenon of the Reformation, and each of these value-laden viewpoints functions as a selector of *some* facts as causally relevant factors.

What is essential is that this form of science's dependence on nonepistemic values — which is undeniable and omnipresent and does not distinguish the human sciences from the natural sciences, despite the fact that the role of nonepistemic values might seem more prominent in the former than in the latter sciences — is clearly *not* sufficient, by itself, to deprive the social or the natural sciences of their value-free character from a *cognitive* point of view. For instance, once we try to explain the causal relevance of economic factors in the historical processes leading to the Reformation, by using *economics* as a selective principle, the causal link we thereby establish is (or is not) valid, according to the evidence it has, independently of our particular economic (or religious) convictions.[5]

Likewise, military interests in projectiles ballistics may have encouraged the birth and growth of modern dynamics at the time of Galileo, but the epistemic warrant of the theory enabling us to calculate the trajectory of projectiles (Newtonian mechanics) does not depend on the different military and political interests. These days, economic, political, and military lobbies do certainly stimulate scientific research by investing large amounts of money in some fields of inquiry while neglecting others. However, the achieved "causal knowledge" (say, the predicted effects of certain chemicals on human beings) does *not* vary according to our nonepistemic preferences and *is* as such *free from nonepistemic values*. We can use our causal knowledge to achieve certain (often morally objectionable) aims; now, if the validity of such knowledge depended on our different nonepistemic values, people with different such values might end up achieving different results by using the same means, and this is obviously not the case.

Consequently, without further arguments at least, we cannot assume that the *cognitive* claims of an empirical, scientific discipline are not objective (that is, intersubjectively valid) simply because, in order to select what is interesting for us, we must restrict our attention to a particular class of phenomena by using potentially nonobjective nonepistemic values.

Insisting on the massive role played by nonepistemic values in the context of discovery enables us to let the possibly *subjective* dimension of nonepistemic values enter the empirical sciences: such a "subjective" dimension of science is, at least in this first context, something to be *encouraged*. It is here that *a plurality of values may favor the growth of science*: in this context, "letting thousands of different flowers bloom" is the right attitude to take. In fact, in both the natural and the historical sciences it is only by looking at the same entity or event from a variety of different perspectives—which presuppose, in turn, many different interests or values—that we can more thoroughly understand the entity or event in question. Just as, following Weber, we should try to understand a historical event from as many different relevant viewpoints as possible—military, economical, political, anthropological, and sociological—a human being can be studied from the biophysical, biochemical, physiological, neurophysiological, psychological, sociological, historical, and gender "perspectives," all of which contribute to a more complete understanding of our multifaceted nature.

In a word, values used as selectors of problems can play their role without necessarily jeopardizing the objectivity of the cognitive claims of science, *provided*, of course, that science *can* and ought to be regarded as an intersubjectively valid enterprise—that is, provided that we can claim that whenever nonepistemic values are the essential factor determining the acceptance of a hypothesis, we are facing an instance of *poor* science or of an *unreliable* piece of knowledge.[6] But since intersubjectivity is one of the epistemic and social values that science is said to promote, I now turn to the second mode in which science and values relate to each other.

2. Values as Selectors of Empirically Equivalent Theories

The second, distinct sense in which (both nonepistemic and epistemic) values could be relevant for science involves the process of comparing and evaluating two rival theories. Epistemic, knowledge-serving values like accuracy, consistency, scope, simplicity, and fruitfulness (see Kuhn 1977, 321–22) have often been regarded as crucial in evaluating two rival theories during a scientific revolution or, more generally, during theory change.

There are two different formulations of the thesis of the empirical equivalence of theories by data. In some overly skeptical renderings of the epistemic power of scientific theories, one could argue that *for any theory T* that has ever been proposed in the history of science, there are (known or unknown) alternative theories T' that are *empirically equivalent* to T.[7] While T and T' are equivalent because they *entail* the same body of evidence, they are incompatible with each other because, for instance, their unobservable substructures explain the observable phenomena by postulating different and incompatible theoretical entities. A less radical but historically more plausible rendering of this thesis claims that *at some moments* in the history of science there are *some* interesting cases of empirically equivalent theories (for this distinction, see Psillos 1999, 167).

Leaving aside the provocative but too often neglected hypothesis that by depending "on an impoverished picture of the ways in which evidence can bear on theories" even the weak form of empirical equivalence of theories might be a figment of armchair philosophers' imaginations,[8] let us assume for the sake of argument that the empirical underdetermination of theories by data (in the less radical sense) is a genuine problem of the methodology of science. On this hypothesis, how should we choose among empirically equivalent theories?

The main problem to be raised in this context is whether purely epistemic values, like those mentioned earlier, are really *sufficient* to resolve cases of underdetermination of theories by the empirical data, so that other, nonepistemic values can always be left *outside* the domain of science (externality versus internality pictures of science's value dependence). As a matter of fact, one of the strongest arguments in favor of science's nonneutrality toward ideological or political values depends not just on the highly controversial thesis of the "empirical equivalence of alternative theories" but also on the observation already made by Thomas Kuhn that his five epistemic values prove "*ambiguous* in application, both individually and collectively," in such a way that they do not determine "a *shared* algorithm of choice" (Kuhn 1977, 331).

To be charitable toward the advocates of a role for nonepistemic values in theory choice, let us suppose that there have been real historical cases in which scientists had to decide between two rival theories, T and T', that at the time were regarded by the experts as being *empirically* equivalent. We could suppose that we could face such a situation even in contemporary science. It is in circumstances like these that epistemic values enter the scene: the *consistency of the two rival theories with other bodies of accepted, well-established knowledge* will obviously become a deciding factor, supposing the unfortunate hypothesis that the two theories are equally well-confirmed by the data, that is, *equally accurate from an experimental or observational viewpoint*.[9]

Now, so the story goes, what should we do in cases where the two rival theories are also both consistent with everything else we know and have the same *scope* or *unifying, explanatory power*? Since *fruitfulness*, or the capacity to generate novel predictions, is very often only a post factum virtue (we cannot know in advance which of the two theories will be "fitter" as shown by its having more "offspring"), we are left with *simplicity*, which, besides its vagueness and language-dependent features, is not a sure-fire guide to truth:[10] ellipses, while less *simple* than circles, still describe the route planets travel!

If we continue supposing the *historically rather implausible* hypothesis that all the "epistemic virtues" (explanatory power, simplicity, fruitfulness, and so on) are satisfied to the same degree by any two empirically equivalent theories, and that the body of evidence remains stable for some time, it would seem plausible to assume that *other, nonepistemic principles of choice should intervene*. If one theory had more acceptable ethical or nonepistemic consequences than the other, should it not be preferred just for this reason? My answer is, of course, that it should! If we could concede all the hypotheses we have introduced so far, in this case an important role for nonepistemic values should certainly be acknowledged.

Note, however, that the way I formulated this example suggests, first of all, that we intuitively distinguish epistemic from nonepistemic values. That we are legitimated to switch from epistemic to nonepistemic values only after our resort to epistemic values fails shows at least two things: in science epistemic values are regarded as hierarchically more important than nonepistemic values, given that we first try to resolve the case of indetermination with the help of epistemic values. Second, we tend to regard the two sorts of values as being *independent of each other*.

After the "wager argument" put forward by Pascal, we know that whenever the *evidential* reasons for two alternative courses of action are approximately equal in weight, we may call into play our *prudential* reasons or, simply put, our nonepistemically based preferences or values. *But I contend that the whole argument by Pascal is predicated on the possibility of separating the evidential (epistemic) from nonevidential or prudential reasons.* Otherwise, Pascal's suggestion of *acting as if we believed in God* in order to end up believing in God would not make sense. It is because we neither believe nor disbelieve at the evidential level that it is necessary to force the body (*il faut plier la machine*) to certain actions, like going to mass or taking the communion. In a word, our prudential reasons for acting (the desirability of an option) should take precedence over our evidential reasons (its probability) only when we have no clear (subjective or objective) evidence for the probability of the conflicting hypotheses. If, in the process of evaluating the expected utility, we calculated the probability only by considering the desirability of a possible state of affairs, our decision would be irrational: believing *in all cases* that our desire for X is going to influence the probability that X will occur is a recipe for disaster![11]

The following example should help us to understand what exactly the issue at stake is in real, as opposed to fictional, science. Suppose we have two physical theories that are empirically and theoretically equivalent (that is, they satisfy each of Kuhn's epistemic values to the same degree) but such that they tell us different stories about, say, the deterministic or indeterministic nature of physical systems: Shouldn't we opt for the theory that pleases us most in terms of the consequences for our freedom (supposing, for the sake of the argument, that either determinism or indeterminism is relevant to make room for a free will)? That this example might not be too far-fetched is clear from the "rivalry" between Bohmian mechanics (which treats quantum probabilities as epistemic) and standardly interpreted, *nonrelativistic* quantum mechanics, according to which the quantum world is irreducibly indeterministic. Shouldn't we endorse the former rather than the latter interpretation of the formalism because of, say, our Spinozian preference for a compatibilist solution to the problem of the relationship between determinism and free will?

Before considering these questions, notice how real-life cases differ from the possible worlds we have been conceiving of so far. First of all, the equivalence of all the *epistemic virtues* of Bohmian and standard quantum mechanics is not to be conceded so easily—for instance, the *explanatory power* of Bohmian mechanics is unquestionably greater than that of the Copenhagen interpretation of quantum mechanics. What matters, of course, is that physicists and philosophers of physics do *not* typically regard such a "theoretical" virtue as explanatory power as sufficient to justify the choice of Bohm's interpretation over Bohr's, for the simple reason that the *empirical* equivalence of the two theories in question is not genuine! Bohmian mechanics *as yet* does not cover *relativistic* quantum mechanics in a satisfactory way (it is not Lorentz covariant), and outside the nonrelativistic regime it cannot be regarded as empirically equivalent to standard, indeterministically interpreted quantum mechanics.

However, for the sake of argument, suppose that *at the present stage of our inquiry* these two physical theories are completely equivalent from *both* an empirical and a theoretical viewpoint. What should we do with our nonepistemic values? Two answers are possible, depending on whether such an empirical equivalence in question can be regarded as *temporary* or *definitive*.

First answer: if the empirical equivalence of the two theories ends up being a *temporary* predicament, that is, a situation that can be overcome by further developing our research, *then* one's commitment to compatibilism or incompatibilism between free will and determinism could provide an excellent reason to *pursue* one theory rather than the other. However, such a reason would clearly be *extraneous* or *external* to the evidence for a physical theory rather than the other, in the sense that it could not justify our belief in the determinism or indeterminism of quantum mechanics as such—unless and until, for instance, Bohmian mechanics were pursued and developed to a point where it would become experimentally more accurate than its rival in all known domains of application. Lacking any empirical progress of this kind, however, I do not see why one should *choose* at all costs between the two equivalent theories by invoking nonepistemic, ethical or metaphysical values: from an epistemic point of view, one should conclude with Galileo's two little words, *non sappiamo*! (we do not know), words that, he remarked, we have so many resistances to pronouncing. As it will become clearer in the next section, I consider this skeptical attitude to be a fundamental *norm* aimed at saving the integrity of science's claim to knowledge. I am not claiming, of course, that there have never been cases in which nonepistemic values have intruded into decisions among temporarily empirically equivalent theories; I am just claiming that the role of these nonepistemic values in science is to push us to know more.

The same remark applies, I dare say, to any other commitment to nonepistemic values. Whenever our commitment to purely epistemic values puts us in a condition of having to face two empirically equivalent theories, we can only hope that temporarily "adopting" or "pursuing" one of the rival theories *even on nonepistemic grounds* can eventually lead us to a situation in which their *empirical* equivalence will be overcome. Accepting a theory on the grounds of nonepistemic values need not imply belief in the theory, which should always involve epistemic values, and in particular experimental accuracy. However, what should we do with a situation in which we suppose that the empirical equivalence is *not* temporary?

This possibility takes us to the second answer to our questions: If the hypothesized empirical equivalence of the two theories were really a matter of *principle*, that is, *if* we could show that we could never know, even in principle, which of the two theories is the correct one, then the situation with respect to our nonepistemic values *could* change. Note, however, that even in this case we would not be *forced* to introduce nonepistemic values to choose between to the two theories, since choosing between them on grounds other than epistemic would be *permitted but not mandatory*. It is only from the point of view of our *nonepistemic* interests that it might be important to choose between the two equivalent theories—this would depend on the theories in question, of course—but it is also essential to recall that *by hypothesis* such a choice would have no effect whatsoever on future empirical research concerning *those* theories, even though it might affect other theories.

From the viewpoint of our *epistemic interests*, in any case, I think it would be plausible to conclude that the difference between two theories that are empirically equivalent in principle on all epistemic virtues is purely *verbal*: the *two* theories are really to be regarded as *one and the same* theory, cast in different but semantically "equivalent descriptions" (Reichenbach 1958, 35; Carnap 1966, 150). Accepting this conclusion does not depend on a neoverificationist theory of meaning: remember that the equivalence of the two theories involves not just experimental virtues like observational accuracy but also more "theoretical" virtues like explanatory power, scope, simplicity, and consistency. If by hypothesis nothing could ever be discovered that would differentiate the theories theoretically and evidentially, one may conclude that a difference that does not make a theoretical and an empirical difference is—at least epistemically speaking—no difference at all. If the cognitive content of the two theories is the same, it is hard even to make sense of the fact that one could choose between them on the basis of nonepistemic values, since, to the extent that we do not have *two* alternative theories, *there is really nothing to choose from*.

Given that it is only by considering issues pertaining to the ultimate aims of science and to the place it should have in our life that nonepistemic values seem to come into play, in order to inquire into whether they should have any such role or not, we should move forward to consider the third type of dependence of science on values.

Before doing so, however, it is appropriate to reemphasize the limited role of nonepistemic values in theory choice and the very remote possibility that two theories can count as being equivalent on *all* the epistemic virtues mentioned earlier. Even more remote is the possibility that the scale weighing the two empirically "equivalent" theories will remain in equilibrium for very long. *There are few if any historical instances of pairs of empirically equivalent theories remaining equivalent for a long time*: the heliocentric theory superseded geocentric astronomy as soon as Newtonian mechanics became established, and Foucault's experiment in 1853 decided the controversy between the wave and the particle theory of light not too long after it was sparked at the beginning of the nineteenth century.[12]

3. Regulative Values of Science and Nonepistemic Values

In considering what I referred to as the regulative, or global, aims of science, namely (1) whether science ought to aim for *truth* or merely for *empirical adequacy* (van Fraassen 1980) and (2) whether it ought to postulate the *existence of unobservable entities to explain* the directly observable phenomena or to be content to describe, predict, and control the observable phenomena by remaining silent about the nonobservable realm, it seems natural to ask whether nonepistemic values play a major role in determining which of these aims scientists typically choose or ought to choose.

Unquestionably, the kind of knowledge science provides is an important value in our life; consequently, trying to understand the function this knowledge plays and ought to play in the overall scheme of things clearly requires considering it in connection with the rest of our values, particularly our *nonepistemic values*. Of utmost importance is *both* how science affects our nonepistemic values *and* whether such values do, and ought to, intervene in scientific methodology.

Schematically, and by recalling Laudan's reticulate model (1984), we can suppose either that (1) nonepistemic values (in the sense previously defined) have an impact on *facts and/or their interpretations* or, less radically, that (2) they affect only the level of science's general values, making us militate *in favor of or against* scientific realism.[13] These two possibilities will be discussed in turn.

Consider a classic example that illustrates how one's commitment to certain nonepistemic values may affect one's methodological choices and eventually have repercussions on the interpretations of facts. Historically, it is not implausible to suppose that opposition to Galileo's realism about the hypothesis of terrestrial motion was motivated by the Church's apologetic attempts to defend both its own authority as the only institution entitled to a correct interpretation of the Bible and an anthropocentric worldview that was in better accord with creationism and therefore with the possibility of keeping some sort of moral and political authority over other human beings. Consequently, the political and ideological character of these goals would make Cardinal Roberto Bellarmino's *instrumentalism* about the Copernican hypothesis a direct consequence of a prior adoption of some nonepistemic values. Bellarmino's belief in Ptolemaic astronomy as the true hypothesis "corresponding" to the facts was compatible with his concession that using the Copernican hypothesis as a mere instrumental device could be useful to simplify the astronomical calculations. For him, there was no evidence at all in favor of the motion of the earth.

Clearly, in this case the interpretation of a scientific hypothesis (the

Copernican one) as a mere algorithmic device affects what should count as "fact," and it is constrained by one's allegiance to some nonepistemic values. Furthermore, if we take for granted that the Church's general political goals included those listed in the previous paragraph, such an instrumentalism concerning the heliocentric hypothesis, this move was not irrational with respect to those goals. In general, and for obvious reasons, treating *all* scientific theories as mere predicting devices and denying them any truth or explanatory capacity are powerful ways to avoid possible conflicts with the teachings of various churches, from the divine origin of all life on earth to the immortality of the human soul.

Note, however, that the strength or degree of plausibility of philosophico-scientific arguments in favor of realistic *versus* instrumentalistic interpretations of science *ought* not be decided on the basis of one's allegiance to atheism rather than to a revealed religion. We believe that good philosophical arguments in favor of or against scientific realism should be acceptable to people of all religious beliefs, atheists included: for instance, the so-called pessimistic meta-induction proposed by Laudan is not an exception to this rule. In short, the reasons we may have for *wishing* a philosophical thesis to be true are not among the reasons we have for *believing* it.

I take it as a fact that since the time of the epistolary between Leibniz and Clarke,[14] the weight of politico-theological arguments (correspondingly, of *some* nonepistemic values) within physics has progressively dwindled. Nowadays, we do not believe that the validity of scientific-philosophical arguments in favor of the absolute or relative character of motion ought to depend, in principle, on reasons external to experimental and/or mathematical physics proper (and therefore on whether it would be better for us to live in a universe where God intervenes often or never intervenes, as Clarke and Leibniz, respectively, believed). This change in attitude toward the role of theological hypotheses within physics from the eighteenth century to the present is due not only to the unverifiable character of such hypotheses but also to the shared, *normative* intuition that theological values may hinder the intersubjectivity of science, given that they are *not* universally shared.

The same argument applies, I surmise, to the rest of our ideological and political values: letting idiosyncratic nonepistemic values play a role in science would run the concrete risk of jeopardizing one of the few cultural conquests of humanity (possibly the only one) that is capable of promoting agreement and consensus, namely science. And, of course, this is one of the nonepistemic reasons why one ought to believe in science, but does not and should not play a role in evaluating single scientific hypothesis. Textbook mathematics, physics, chemistry, biology, psychology, and economics are examples of knowledge *de facto* shared across different cultures and ethical and religious values. If science's claims to knowledge depended on such nonepistemic values, people of different nations would not agree on textbook science as they in fact do.

Of course, this is not to deny the existence and importance of debates within science, but they are crucially directed toward areas of less reliable knowledge, and they should not be resolved in terms of one's allegiance to nonepistemic values. The agreement in science is explained by scientists' *shared* commitment to epistemic values like consistency, experimental accuracy, and scope.

Whenever our pursuing epistemic values seem to have an impact on our nonepistemic values, are we allowed to maintain that facts totally depend on nonepistemic value-relative interpretations?

It may be true that the very strong evidence in favor of Darwinian evolutionism has the effect of weakening one's faith in Creationism, but this does not mean that the evidential relationships linking facts and hypotheses in evolutionary biology depend on some biologists' atheistic or materialistic beliefs, as some Italian intellectuals have recently maintained. Note that if these intellectuals were correct, we should introduce Creationism in high schools in the name of a *pluralistic* stance on *ethical* values, and this obviously does not seem right: Creationist and evolutionistic explanations of the origin of our species can and should be compared purely on the level of evidence-theory relationships.

In a word, if we defend the value-ladenness of scientific facts in this more internal, constitutive sense (not in the external sense according to which facts are only selected by our values, as in section 1), I see no argument to counter a neo-Nazi's claiming that the "alleged" facts concerning the Holocaust are constituted by the nonepistemic values of the Jewish oligarchy. If we "ideologize" the natural and historical sciences through and through, the price is letting our will to power subjugate every domain of our experience. The view that *all* natural and historical facts are constituted in a strong sense by our nonepistemic values may engender in those who hold it a feeling of self-importance and power, but, besides being based on a wrong philosophical argument, this view may significantly reduce the possibility of a peaceful coexistence of our species. And this is another nonepistemic value that should make us opt for the kind of knowledge provided by science, even though such a value never intervenes in adjudicating internal scientific disputes. The wrong philosophical argument depends on the simple mistake of not recognizing that any possible inquiry into the natural world *presupposes our assuming that there is a way things are*, that there are *facts* that are independent of our wishes and cognitive or noncognitive preferences. Without this assumption of independence of facts from our nonepistemic values, trying to find out how things are would be a meaningless enterprise.[15]

Note, furthermore, that a relevant amount of scientific knowledge seems to be completely devoid of any social consequence whatsoever. Prima facie at least, the fact that planets have elliptical orbits around the sun or that hydrogen is more abundant than helium in the universe has no relationship to our being committed to racism, sexism, feminism, or any other political or ideological value. Even if Hume's rule "no ought from is" were false (a big if), it is not at all clear how one could use facts like these to promote one's ideological case. Not all science is so disconnected from our ethical and political interests, of course, and one ought to be careful in denouncing ideological uses of science. But all such uses are either (1) a violation of Hume's rule, or (2) an illegitimate attempt at using a nonepistemic value as the only reason to believe in an empirical hypothesis, or (3) make reference to cases in which our knowledge is very unreliable.[16]

It might be surmised that the defender of a role for nonepistemic values in the third sense need not go so far as to deny the independence of facts from nonepistemic values and be content with the remark that the aims of science in general are and should be intrinsically constituted by our nonepistemic values. Given the undesirable consequences of denying the existence of value-neutral facts in the relevant (internal) sense, arguing that our stance about the issue of scientific realism is infected by our nonepistemic values would then appear to be a much more plausible road to take.

To understand this more moderate claim, consider a beautiful example taken from Albert Einstein's autobiography (1949), in which we are told that the natural sciences, interpreted as an effort to decipher a mind-independent, external world, can be ascribed the overarching purpose of "freeing us from the chains of the merely-personal":

> It is quite clear to me that the religious paradise of youth, which was thus lost, was a first attempt to free myself from the chains of the "merely-personal," from an existence which is dominated by wishes, hopes, and primitive feelings. Out yonder there was this huge world, which exists independently of us human beings and which stands before us like a great, eternal riddle, at least partially accessible to our inspection and thinking. The contemplation of this world beckoned like a liberation, and I soon noticed that many a man whom I had learned to esteem and to admire had found inner freedom and security in devoted occupation with it. . . . The road to this paradise was not as comfortable and alluring as the road to the religious paradise; but it has proved itself as trustworthy, and I have never regretted having chosen it. (Einstein 1949, 4)

It could be maintained that a belief in a mind-independent "external" reality was an essential presupposition and motivation of Einstein's research, and that, plausibly, a belief in unobservable, mind-independent entities was part and parcel of his religious appreciation for the mysteries of nature. Believing that what exists coincides with the bound of our perceptions—like the typical idealistic philosopher would have it—would not have helped Einstein to free himself of the chains of the purely personal or to relate his limited spatio-temporal existence with the observable universe as a whole. And yet the EPR argument[17] in favor of the incompleteness of quantum physics and therefore of a more realist understanding of the theory, though possibly *motivated by Einstein's idiosyncratic religious faith*, obviously did not make any reference to his "religious" faith in realism. Einstein's argument against the view that quantum mechanics as Bohr interpreted it had to be regarded as complete did not mention his allegiance to scientific realism or his belief in the independence of the quantum world from our measurements but tried to uncover technical difficulties that a physicist of different philosophical convictions like Bohr could acknowledge and worry about.

Likewise, in claiming that the molecules postulated in statistical mechanical models are real, Ludwig Boltzmann was surely motivated by a *general* realistic attitude toward science that he defended in various ways in his scientific and philosophical production. However, until a *common-cause argument* could be produced in favor of the atomic theory of matter, one that made reference to thirteen different empirical methods to calculate Avogadro's number, all converging to the same value, Boltzmann's belief in the reality of molecules was not able to conquer the vast majority of his colleagues.[18]

The moral of these examples is twofold. First, general philosophical convictions about the goals of science are usually *not* sufficient to conquer scientists' allegiance to a *particular* interpretation of a *particular* scientific theory, since typically only sufficient experimental or mathematical work can decide the issue. And even where the issues are still unsettled, as is often the case in philosophy, the commitment to certain nonepistemic values can at most *motivate* a philosopher to build an argument, but such values by themselves are not sufficient to regard the argument as good. Consider, for example, van Fraassen's criticism to the inference to the best explanation, sometimes used by scientific realists to claim that the truth of science is the only explanation that does not turn its predictive success into a miracle (van Fraassen 1980). The validity, plausibility, or strength of van Fraassen's criticism do not depend on his religious motivations to defend constructive empiricism, and also a scientific realist can recognize that inferences to the best explanation are not sound arguments, if indeed they are not.

The second moral is that perhaps we should stop worrying about "the general aims of science," because we can give a clear answer to the question of correctness of the ontological assumptions of science only by looking at specific, particular scientific theories. Depending on the arguments at hand and the degree of maturity of a certain theory, one can be a scientific realist about a particular set of hypotheses while accepting an instumentalist stance toward a different set.

Finally, notice that whether truth is to be regarded as the aim of science is a de jure question, given that most scientists de facto regard the discovery of truths about a mind-independent, unobservable world as the main goal of their efforts. Philosophers of course debate whether scientists are justified in doing so, that is, whether the epistemic and methodological resources (*the means*) yielded by current scientific doctrines are adequate to reach truth (*the end*). Since all means-to-ends relationships in principle are descriptive and subject to empirical inquiry (the means can be regarded as "the causes" of the attainment of our aims, which are their effects), the question of truth in science can be analyzed without resolving the normative problem, since it is amenable to rational and objective discussions as is the correctness of any other hypothetical imperative of the kind, "if you want x, do y."

4. Applied Science and Nonepistemic Values

Besides their role in choosing facts and problems, as discussed earlier, *nonepistemic* values also play a fundamental role in the application of our scientific knowledge to policy making. The structure of the argument is simple and well-known: whenever the *acquisition* of new knowledge or the technological *application* of already possessed scientific knowledge have nonepistemic consequences, say, on the environment, our health, or economic production, nonepistemic values are involved. Often, such values take precedence over the epistemic ones, sometimes for good reasons, sometime for bad. For example, if scientists wanted to find out (an epistemic aim) whether a certain chemical that would have wide industrial use increases the risk of human beings getting cancer, they should not be allowed to experiment directly on people. In general, whenever the application of already possessed knowledge, the means to achieve new knowledge, or the production of new technological devices have damaging nonepistemic (ethical or environmental) effects, we are obliged to refrain from applying our knowledge or from satisfying our epistemic curiosities.

In addition to these noncontroversial considerations, it has recently been argued that there are cases in which nonepistemic values "have a legitimate role to play in the *internal* stages of sciences" and "are required for good [scientific] reasoning" (Douglas 2000, 565, emphasis added). In what follows, I will briefly evaluate Douglas' claim that the so-called inductive risks carried by many statistical hypotheses pose a threat to the standard view of scientific reasoning as value-free in the "internal sense" explained earlier.

Inductive risk, as Hempel defined it, is the chance that one's hypothesis is false, despite its having been accepted, or the chance that it is true, despite its having been rejected (1965, 92). In a word, inductive risk is the (sometimes nonnegligible) chance that our scientifically acquired beliefs may be wrong. Douglas considers the interesting methodological problem of having to set a standard for the *statistical significance* of certain toxicological tests. While *stricter* standards of statistical significance (in comparison with control groups) will reduce the number of false positives and increase the number of false negatives (making the chemical appear to be *less* dangerous than it actually is), *laxer* standards will increase the number of false positives as well as increase our chance of considering ill an animal who is healthy: in this case, the chemical will appear to be more dangerous than it actually is. The question that arises is, *if we cannot lower the number of both false positives and false negatives, what should we do?*

In cases like these — even before applying our knowledge coming from the result of the tests, and consequently deciding to prohibit or allow the production of a certain toxic substance — it would seem, so Douglas argues, that the "scientific," empirical claim itself depends on a previous choice of *overregulating* (by setting lower standards of significance) or *underregulating* (by setting higher standards of significance). Overregulating may better protect our health but may have a greater cost for the economy; underregulating may be more dangerous for us but perhaps costs less for the industries involved. In any case, both decisions seem to be strictly dependent on *nonepistemic* values.

It could be objected that, quite independent of one's preference for certain nonepistemic values over others — whether to be more protective of health or a manufacturer's interests — tests forcing us to arbitrarily balance false positives and false negatives exemplify *bad scientific tests* or *bad science*. Here "bad science" and "bad tests" simply mean science and tests that are not conducive to *highly reliable* results, unlike other ordinary tests that are used to diagnose the presence of certain diseases, such as those for prostate cancer. All we should do to improve the situation from an epistemic point of view is to make our sample larger or to change the test in some way, so as to lower the inductive risk.

This desirable strategy, however, may not always be possible. Consequently, in order to find the toxicity level of the chemical, some significance level must be set, and such levels must be determined, as Douglas's case is set up, by the two nonepistemic values of protecting our health or favoring the production of useful but toxic chemicals. And while such tests are not as reliable as other tests in other areas, we still need a policy in the short run, since "in the long run we'll all be dead," as Keynes put it, so we cannot wait around to see if other, more reliable tests will be found some day. As Douglas puts it, "some balance must be struck" between the need for reducing the number of false positives and the need for reducing false negatives — needs that for each particular kind of test cannot be satisfied simultaneously.

Douglas concludes that in cases like these, nonepistemic values *determine to some extent* the empirical content of our hypotheses. While we are obviously aware that this is the case, we somehow "use our ignorance" to act on the basis of one nonepistemic value rather than another. And despite many of us feeling that setting laxer standards of significance is preferable in order to protect our health, we all know that economic interests at times may be so powerful as to succeed in convincing the "experts" to choose otherwise.

Independent of what occurs in real life in cases like this, it seems undeniable that Douglas's and analogous arguments in favor of the intervention of nonepistemic values in the so-called internal (that is, evidential knowledge-gathering) processes are predicated on the existence of *still uncertain areas of scientific inquiry*—that is, areas in which we still do not know the real effects of certain kinds of causes. *It is exactly this lack of reliable scientific knowledge that explains the interventions of nonepistemic values in the setting of parameters for the significance of the test.* Douglas seems to accept that *ignorance* would be a better word for what concerns the toxic power of dioxin, the case she discusses: "Where the balance should lie for dioxin studies is currently unclear" (2000, 569).

Those who are in charge of deciding what to do should keep this conclusion in mind, since claiming that we have reliable statistical knowledge and are therefore entitled to deregulate or overregulate according to the case is one thing; it is quite another to openly admit our ignorance. Fortunately, not all cases of applied science are like this—think of all the situations in which a *positive* result in a test uncontestably indicates the presence of the disease (consider diseases such as AIDS, epilepsy, and some cancers)—but it must be admitted that a certain amount of inductive risk accompanies any decision in policy making.

Note, however, that the methodological problems of setting the parameters for the significance of the test do not jeopardize at all the *intersubjective validity* of the empirical claim made by the statistical hypothesis. In fact, regardless of the standard (strict or lax) one chooses according to one's preferred nonepistemic values, both parties—the overregulators and the underregulators—agree that in both cases our statistical hypotheses carry a measurable margin of error. Namely, they agree that *by choosing a stricter standard dioxin will appear to be less dangerous than it is, and by choosing a laxer standard dioxin will appear to be more dangerous.*

If this is the content of the cognitive claim before we decide which test to administer, we have to try to clarify the sense in which the content of the hypothesis is biased by the choice of the two nonepistemic values. As explained in section 1, we often need and use nonepistemic values to select facts or empirical hypotheses, or to stress one causal factor rather than another in a very complex causal field. Is there any significant difference between the examples discussed in section 1 — in which two different nonepistemic values selected different facts or different relevant causal aspects of the same fact — and the case discussed by Douglas?

In Douglas's case, the choice of two different nonepistemic values make us evaluate in a different way the probability of an empirical hypothesis (namely, that dioxin causes cancer), so that the empirical content of the hypothesis, the assigned probability, seems to depend on the nonepistemic value. This claim is compatible with the fact that the empirical hypotheses made in the two cases as a result of the two alternative choices are objectively valid, because *both* the underregulator and the overregulator *agree* that by choosing the standard of significance in different ways the dioxin will appear either less or more dangerous than it actually is. Compare this situation with a disagreement about relevant explanatory factors concerning the importance or relevance of a particular causal factor, depending on whether, in order to explain a car accident, for example, one privileges the speed of the car, the ice on the street, the bad brakes of the brand new car, or the slow reflexes of the driver. The dependence of causal explanations on pragmatic factors presents similarity with the case Douglas discussed, since the car's insurer, manufacturer, and driver may want to stress some factors over others. Also, in this case, if no more information is available to decide whether one causal factor weighs more than another (provided all are present), it is not irrational to introduce one's nonepistemic preferences, but this proves nothing about the subjectivity of scientific knowledge. The problem of having to act or decide despite inductively insufficient evidence was of course known long before the case discussed by Douglas. In all cases where we must evaluate the probability of a course of action and have no objective frequencies (chance) to rely on we must trust *subjective probabilities*, in which desires and beliefs, prudential reasons and evidential reasons, may be inextricably meshed. For example, to have a good chance of successfully passing an examination, it may be rational to lead ourselves to believe that we will, because, as William James taught in similar cases, *our chance of passing the exam, depends on our believing that we can act in such a way as to pass it*. Since the test of whether such a belief was justified can only be given by our taking the exam, in all cases in which we lack sufficient evidence and the desired outcome depends in part on us, it is rational to open the door to our nonepistemic values and interests.

In short, when we have "to strike the balance" for the statistical significance in the dioxin studies, we are just looking at another manifestation of Pascal's wager: "Act as if God existed, if this uncertain state of affair is highly preferable to you, *provided that* the evidence for its existence equals that for its non-existence and the corresponding expected utility is greater." However, whenever *prudential* reasons can be sharply separated from *evidential* reasons — namely, when our empirical knowledge is highly reliable and we cannot just believe what we like — I cannot see how inductive risks in Douglas's sense pose any threat to Longino's claim (1990, 85–86) that values intervene in the *application* of scientific knowledge and not in its *constitution*. Douglas seems to agree on this point, since she recognizes that "Hempel was right in asserting that whether or not a piece of evidence is confirmatory of a hypothesis ... is a relationship in which value judgments have no role" (2000, 564).

To summarize, Douglas's dioxin case appears to favor the view that nonepistemic values intervene in the constitution of the evidential claims of science only because *we still do not know how things really are*. But since Douglas is certainly not prepared to argue that we are always as ignorant as we are in the dioxin case, the value of her analysis in cases where the application of our scientific knowledge has social impact is limited to cases in which our hypotheses are as unreliable as those involving the dioxin case. Furthermore, if we consider that in *many* areas of astrophysics and theoretical physics and other bits of natural science no social impact is foreseeable and no nonepistemic values seem to intervene in the formation of the hypotheses, the impact of nonepistemic value in evidence-theory relationships should not be exaggerated (see Douglas 2000, 577, for her admission of this point).

Certainly, if in the dioxin case our knowledge were more reliable, the balance between false positive and false negative would not be struck according to one's evaluation of the relative importance of ethical values and economic constraints. Since deciding on the basis of the available tests may be leading us into an irresoluble conflict of nonepistemic values, the moral of the story is that epistemic values should take precedence whenever possible, not by chance. Also, Douglas admits that we should try to *know more* by changing the experimental technique or by increasing the number of tested cases (Douglas 2000, 566).

Conclusion

The role of nonepistemic values in science, as it relates to the first three cases in which science depends on values, seems to be limited to the context of *pursuing* a theory, while in the fourth case it is linked either to the various necessities of applying our knowledge or to formulating belief partially supported by nonepistemic values in order to reach a decision. In all four cases, however, no serious objection to the neutrality of the cognitive content of scientific hypotheses seems forthcoming, and this is the result that I find most important. The fact that nonepistemic values may influence the methodology through which the empirical results can be gathered in cases in which we still do not know much about the empirical links between causes and effects is just an indication that we have unreliable knowledge, and, in this sense, no "good science" at all.

Much more interesting would be the claim that social, nonepistemic values *ought to* condition our acceptance of a theory from an evidential point of view, namely, that we should believe in a certain hypothesis because it supports one ideological position rather than another; however, no serious attack in this direction seems in the offing from the type of cases Douglas studied. That nonepistemic values sometimes do condition the acceptance of a theory is, of course, true, but this is a *bad* thing: science *ought* to steer clear of nonepistemic evaluations of the evidential support of hypotheses or theories. Whenever attacking the intersubjective validity of the theory-evidence relationship threatens to become a favorite sport, we should remember that if the *existence* and the *explanation* of facts depended purely on one's involvement with certain religious, ideological, or political values, only the economically, religiously, and politically powerful would have a chance "to be right."

NOTES

I thank Fabio Bacchini, Heather Douglas, Peter Machamer, and Gereon Wolters for their numerous critical comments to an early draft of this chapter.

1. For instance, according the type of relationship linking hypothesis and evidence, the former could qualify as plausible, probable, or certain, typical evaluations we attach to hypotheses. Consequently, "certainty," "high probability," and "plausibility" count as epistemic values.

2. Of course, this claim is consistent with the fact that the hierarchy among these values is different for different scientists.

3. This is because of the Humean separation of *is* from *ought*.

4. We will see that it is the social control ensured by the public character of science and the *actual* commitment of scientists to values such as experimental/observational accuracy of hypotheses and their logical consistency that de facto ensures science's intersubjective validity.

5. For Weber's famous *Wertfreiheit* as distinguished from the *Wertbeziehung*, see Weber (1904). For a more recent evaluation of this problem that defends the same conclusion, see Longino (1990, 85–86).

6. Whether science can reach objectivity, in the sense given by the discovery of a mind-independent world, is a different problem but strictly related to the possibility of explaining the kind of intersubjective agreement that we reach in the sciences. I investigate this other sense of "objective" in part in section 3.

7. Philosophers that, like Quine, have stressed the paucity of inputs (empirical data) vis-à-vis the richness of the hypotheses that are produced to explain them (outputs) may belong to this category. But in this context I am interested not in capturing anyone's position but merely in exploring a possibility.

8. See Norton (1993, 1), whose analysis is a brilliant defense of this thesis, illustrated with the case study of the origin of quantum discontinuity.

9. To be charitable toward the defender of a role for nonepistemic values in theory choice, I am supposing, against Laudan and Leplin (1991), that empirically equivalent theories are not differentially confirmed. For a criticism of Laudan and Leplin, see Okasha (1997).

10. Algorithmic complexity theory may significantly improve our understanding of the requisite for "simplicity" in a theory, provided that the theory is axiomatizable. In this case simplicity can be measured in terms of the length of the shortest program that is necessary to generate the theorems (laws) of the two theories.

11. An important exception to this rule will be discussed in the fourth section and has to do with cases in which the occurrence of the future state of affairs depends on our belief that the state of affair itself will occur.

12. Stathis Psillos (1999, 168) also stresses these points.

13. In Laudan's view the methodology of science can be construed as a set of hypothetical imperatives of the type, "if you want aim x, follow methodological rule y," where x represent our cognitive and (possibly) noncognitive aims. Among such cognitive aims, here I focus on those belonging to the issue scientific realism versus antirealism.

14. Their correspondence begins in 1715. The epistolary is collected in Alexander (1956).

15. For this transcendental argument, see Searle (1998).

16. The third disjoint of this alternative will be discussed in the next section (section 4).

17. This argument used quantum correlations to argue that quantum mechanics is either nonlocal or incomplete. Using two particles measured in different locations in space but originating from the same source, one could predict with certainty, according to Einstein, Podolsky, and Rosen, the measurement result on the distant particle after having measured the particle nearby. If measuring the particle in one wing does not disturb the distant particle (*locality*), the fact that one can predict with certainty the far-away result entails that the particle had a definite value before the measure, which means that the theory as standardly construed is incomplete.

18. The classic text in this context is Perrin (1913). The common-cause structure of the argument has been clarified by Salmon (1990).

REFERENCES

Alexander, H. G., ed. 1956. *The Leibniz-Clarke correspondence*. Manchester: Manchester University Press.
Carnap, R. 1966. *Philosophical foundations of physics*. New York: Basic Books.
Douglas, H. 2000. Inductive risks and values in science. *Philosophy of Science* 67 (4):559–79.
Einstein A. 1949. *Albert Einstein, philosopher-scientist*, ed. Paul Arthur Schilpp. New York: Harper & Row.
Hempel, C. 1965. Science and human values. In *Aspects of scientific explanation and other essays in the philosophy of science*, 81–96. New York: Free Press.
Kuhn, T. 1977. *The essential tension*. Chicago, IL: University of Chicago Press.
Laudan, L. 1984. *Science and values*. Berkeley: University of California Press
———. 1996. *Beyond positivism and relativism*. Boulder, CO: Westview Press.
Laudan, L., and Leplin J. 1991. Empirical equivalence and underdetermination. *Journal of Philosophy* 88:449–72.
Longino, H. 1990. *Science as social knowledge: Values and objectivity in scientific inquiry*. Princeton, NJ: Princeton University Press.
Norton, J. 1993. The determination of theory by evidence: The case for quantum discontinuity. *Synthese* 97:1–31.
Okasha, S. 1997. Laudan and Leplin on empirical equivalence. *British Journal for the Philosophy of Science* 48 (2):251–56.
Perrin, J. 1913. *Les atomes*. Paris: Alcan.
Psillos, S. 1999. *Scientific realism: How science tracks truth*. London: Routledge.
Reichenbach, H. 1958. *The philosophy of space and time*. New York: Dover.
Salmon, W. 1990. *Forty years of scientific explanations*. Minneapolis: University of Minnesota Press.
Searle J. 1998. *Mind, language, and society*. New York: Basic Books.
van Fraassen, B. 1980. *The scientific image*. Oxford: Oxford University Press.
Weber, M. 1904. Die Objektivität Sozialwissenschaftlicher und Sozialpolitischer Erkenntnis. *Archiv für Sozialpolitik und Sozialwissenschaft* 19:22–87.

4

The Social in the Epistemic

Peter Machamer
Department of History and Philosophy of Science, University of Pittsburgh

Lisa Osbeck
Department of Psychology, State University of West Georgia

In considering the role of values in science, it has been traditional to acknowledge a distinction between two principal categories or kinds of values: social values and epistemic or cognitive values. Like many things taken for granted, however, the basis for this conventional separation appears fragile when closely examined and its force and import are deserving of considerable rethinking. Various possibilities exist for renovation. One might attempt to strengthen the basis of the distinction by bringing greater clarity and precision to it or by drawing finer differentiations on one side or the other (as Laudan does in chapter 1, for example, by distinguishing "the cognitive" from "the epistemic"). Yet any move of this kind would still maintain the independence of the two (or three) kinds of values. An approach we find preferable to either of these is to deny the independence of epistemic and social values and, for that matter, the distinction between so-called personal values and social values. The ultimate project for describing how values work in science should be to clarify how and where values enter into the doing of good science, and how and where they can make for bad science. Within this project it will be important to spell out where and when values promote bad science, but there is little practical or theoretical worth in demarcating these by kind, that is, as social *or* epistemic. Here, we will lay the groundwork for the project by showing that there is at most one way to draw a simplistic social and epistemic distinction, and that such a strategy would not satisfy most theorists as a basis for describing the kinds of values that enter into doing good science. Of course, from here it follows that using an epistemic/social distinction as the way of distinguishing good science from bad is even more heinous. Less dramatically, it means that even those who would involve social values in the procedures of good science ought not do so in ways that depend on a bifurcation from the cognitive. Fundamentally, we argue that epistemic, or cognitive, values are ineliminably social, and that this is so in many important ways.

Let us be clear about our claim. No one doubts that we can separate the algorithmic calculatory rules as found in Bayes's theorem, error theory, TETRAD, or formal learning theory from other heuristic or informal "rules" used for evaluating or testing theories. These algorithmic rules apply only to a formalized theory-evidence relation. In fact, the theory-evidence relation is generally the locus for most claims that epistemic values ought not to be affected by the social. That is, it is held that theory-evidence relations ought to be purely epistemic and not contaminated by any social or nonepistemic factors.

C. G. Hempel, in his classic analysis "Science and Human Values" (1965 [1960]), maintained that values might enter scientific knowledge production at any point *except* in the "logical" theory-evidence relation. That relation was determinate and fixed by some calculus of inductive or statistical support. His thought seems to be that if we had such a calculus of confirmation, or some such algorithmic relation, we would put theory-evidence relation inferences or reasoning beyond the pale of intrusive values that could sully the purity of that epistemic connection. Yet even Hempel did not think that these evidential relations determined theory choice or theory acceptance; the extent of evidence considered sufficient to accept a theory was open to influence on many other grounds. Values were in principle and practice intrusive in the acceptance of theories, and they could intrude, presumably, in good or bad ways.

If the category *epistemic* is limited to only these evidential algorithmic rules of support, then many people might agree with Hempel and claim that other nonepistemic values have a role only in hypothesis or theory acceptance and in theory evaluation. Larry Laudan and Hugh Lacey (chapters 1 and 2, respectively) seem to be clearly in the Hempelian camp, as is Heather Douglas (2000). Laudan, for example, argues that the epistemic as construed above as calculatory is not sufficient for theory acceptance or evaluation. We need rules about things like simplicity, fecundity, and scope: values that are more accurately and broadly called "cognitive." Most everyone agrees on the need for such values, that is, those that transcend the algorithmic, and that they play a role in theory acceptance and so in good science. Yet many, including Lacey and Laudan, still hold that they remain distinct from other nonepistemic, noncognitive values, that is, from values that are social.

However, here are some questions. What distinctions may be drawn among the kinds of values that can legitimately enter into doing "good" science? Within that class of values, can there be some significant mark of independence between the cognitive and the social? Are these really distinct kinds of values? Moreover, if they are, would it then suffice to demarcate social values so that they may be ruled out, and in doing so prevent philosophers (and scientists, too, presumably) from falling into epistemic error? This would still leave open the possibility that in some cases social values could enter science in morally apt ways; they just would have no epistemic role.

We are not convinced that the social and the cognitive are distinct. In fact, we have yet to find any clear account of the cognitive or epistemic that transcends the algorithmic. Truisms and vagueness abound. Examples or lists are offered in place of definitions or explanations, and still the basis for the cognitive-epistemic distinction does not emerge clearly. The lists of epistemic-cognitive values from Merton, Kuhn, Laudan, and now Lacey are well known, including such items as simplicity, fecundity, impartiality, and avoiding bias. But as Kuhn (1973) points out, there is room for much individuality and social constraint even in the application of these cognitive "rules." Barry Barnes makes this point in chapter 5.

So to uphold the significance of the distinction, we need a principled way by which these cognitive values *and* their "rules" of application could be cleansed of the personal and social elements, making them pure epistemic rules indeed. But here is another, perhaps more important point: No one has specified what the *social* includes that ought to be (let alone can be) excluded from the cognitive such that the cognitive constituent might be independent.

Here our concerns are shared by Longino (2001; chapter 7), who has emphasized that the dichotomous understanding of social and cognitive values by those on both sides of the debate has led to an unhelpful impasse; she has also pointed to the lack of an "in principle way mechanically to eliminate social values and interests from such (scientific) judgment" (chapter 7). Longino's approach emphasizes the inadequacy of purely logical or cognitive criteria, the "gap" in justification that logic leaves. She advocates, then, an "expanded justification," which would include the social dimensions of cognitive activity in the practice of science, particularly in the form of public scrutiny and criticism from a variety of perspectives.

While we appreciate her efforts to include the social in the process of justification, and agree somewhat with her claims about perspectivalism, we would not exclude the theory-evidence relation from being social in some aspects, nor do we think that the best way to proceed lies in "adding to" the cognitive values that are already there. This supplementary strategy assumes in a subtle way that it is possible in principle to separate social from cognitive dimensions, such that the social might then be used to critically assess the cognitive. Instead, we take the approach that it is fruitless, even misguided, to attempt to delineate independent epistemic or cognitive aspects from the social (or maintain the independence of epistemic from social values). We shall argue that the social is constitutively part of the cognitive (and vice versa). Values (or constituents of values) traditionally considered social must, in principle and practice, be part of any analysis that purports to explain both the functioning and normative force of cognitive and epistemic values.

The Natures of the Social

Longino (2001) rightly notes that the ambiguity of the word *social* adds to the muddle. Let us start by trying to outline the several senses of *social* that seem to undercut any claim for the independence of knowledge or epistemic values. We shall call these basic senses acquisition, memory, use and norms, and demonstration. These are loci where the social determines, causally and normatively, scientific knowledge.

Acquisition

The first sense of *social* manifests itself in the processes through which learning occurs and knowledge is acquired and structured. A prominent version of this has been called the "sociogenetic" position on mind (or knowledge and mental processing), a position gaining ground in psychological and social science circles over the past thirty years and foreshadowed by such thinkers as L. S. Vygotsky, George Herbert Mead, Pierre Janet, and Alfred Baldwin (see, for example, Valisner and van der Veer 2000).

One paradigmatic instance of sociogenesis of mind is that we learn to speak from causally integrating and responding to people with whom we share a language and many cultural referents. Further, the intent in using language, from the first, is to communicate with others; and others' responses to our utterances, along with, perhaps, some innate "rules" or constraints, determine what we learn to be meaningful. The most obvious examples of the social nature of learning derive from the uncontested fact that our actions and utterances are assigned meaning and thus rendered sensible within a social context, system, or "form of life." Linguistic categories arising from this communication structure are learned or acquired.

Yet learning and its social nature are not limited to language. Our very conceptual scheme and many of our discriminatory abilities are caused by social interactions. Clearly, in many kinds of learning, the physical world plays a causally effective and important structuring role. The physical and the social and our actions/reactions, as mutually interrelated systems, determine what we take in, how we represent the world, and how we "process" information.

This claim about learning applies no less to acquiring and using scientific knowledge. Acquiring a specialized lexicon is certainly an important part of being initiated into a scientific specialty. However, there is more to the initiation than mastering the lexicon. There are canons of procedure, official instruments for measurement and models of analysis (such as techniques for using microscopes, ways of determining masses, and procedures for running multiple regressions), standards for interpreting and communicating results, all of which are socially established and endure for the historical periods during which they are maintained. These are some of the forms and structures that make possible the acquisition of scientific knowledge. Similarly, one acquires in a social context knowledge of what is valued in theories. One must be taught initially, by others, to recognize simplicity and fecundity in relation to theories, implicitly (by modeling or problem solving) or explicitly.

Laudan (chapter 1) describes this acquisition sense of social as being so broad that it is vacuous. We have argued previously (Machamer and Osbeck 2003) that this sense of the social is appropriately broad, but hardly vacuous, as underscored by the thinkers who, historically, have taken this position seriously and by the different structures of knowledge that result from different learning processes. This sense serves as an important reminder that science is a deeply and inextricably human enterprise, that humans live in "networks of relationships" (Longino, chapter 7), and that humans are social, discursive beings with respect to what they learn. In short, learning is a social process and what is learned has social content.

Memory

It makes little sense to speak of acquiring or learning if what is acquired is not stored in some cognitive system or remembered. Any knowledge must be "represented" in one or more of the memory systems, whether sitting calmly as a trace, as a somewhat permanent systemic modification of some kind (such as a long-term potentiation), or by being reconstructed from stored elements on the occasion of its (conscious or unconscious) recall. Now, most knowledge representations, other than purely idiosyncratic episodic or autobiographical ones (if such are even possible), will have something in common with the memories other people have. Therefore, another sense of the social aspect of knowledge is shared memory, as in discriminatory reactions, concepts, linguistic categories, assumptions, tastes, preferences, and standards people hold in common. Kuhn (1973) noted some of these constituents of group memory. Longino (1990) also notes some that frequently function implicitly as presuppositions within a scientific community, often taken for granted and unarticulated, yet powerful in structuring and directing actions through goals, procedures, problem solving, and even interpretation and observation. In some cases that are most important in science, these shared memories reflect what Valsiner and van der Veer (2000, 5) call "the institutionalizing of a particular knowledge-construction device," for example, establishing the use of statistics in the social sciences such that it becomes the normative, shared goal and methodological cannon of the discipline.

These representational structures are a group's internalized shared schemata, including their conceptual schemes. Yet there is another social dimension to be noticed here. The internal parts of memory (the content of what is known and represented in memory) also include social memories. These may be seemingly trivial, such as remembering which fork to use at a formal dinner, or more epistemic, as in recalling where a co-worker stored the proper form of regression analysis to use when trying to analyze a certain type of data.

Use and Norms

There is an important sense in which identifying something as *knowledge*, whether procedural, declarative, or otherwise "encoded" in memory, presupposes that shared social norms of appropriate or correct *application* hold for what is acquired or known. These application norms pertain to the uses of what has been internalized and function as criteria for whether one has acquired knowledge at all. To be counted as knowledge, what is learned and stored must be used by the knower. It is epistemologically insufficient to theorize knowledge only in terms of acquisition and representation. One must also consider how people *exhibit* learning as having occurred by taking appropriate or correct action (including correctly using language). For example, Philomena might use a fact she has learned as a premise in an argument as she reasons her way toward a conclusion, or she might express this fact by saying it to other people and then go on to draw a conclusion based on it. Someone in the audience, say Fred, hearing her speak, must at least tacitly acknowledge the grammatical correctness of her sentence, and might approve of or criticize the inferences she draws from it. He will further demonstrate his evaluation of her "fact" by using it himself to make a point on a later occasion. In this example both Fred and Philomena are using rules or procedures of grammar and inference that are socially accepted and constitutive of proper grammar or proper reasoning. Normative criteria for knowledge and knowledge claims are social, public, and themselves part of shared memory. What is learned has to enter into the public sphere and be warranted in public ways in order to count as knowledge. Traditionally, learning has been either defined or, certainly, tested in terms of performance (say, how a student performs on a science exam). This is surely Wittgenstein's point (1953) in tying acquisition of linguistic rules (or more generally, grammar) to socially appropriate use of language and the following of these implicit rules.

Knowledge is constituted, in part, by action, broadly defined — action here includes appropriate judgment and use of words. And it is inseparably so constituted; that is, knowledge always has a constitutive practical part. We might say that every bit of theoretical knowledge is tied inextricably with practical reason, a thesis that dates back to Aristotle (see, for example, *Nicomachean Ethics,* VI, 7,1141b13ff.). Importantly, these knowledge acts occur in a social space wherein their application is publicly judged as appropriate or adequate (or even correct). One way to see this is to note that truth conditions, at least in practice, are use conditions, and even those "norms set by the world" (empirical constraints) structure human practices concerning correct or acceptable use, for example, of descriptions of events, use of instruments, or even the right ways to fix a car. Human practices are social practices, and the way we humans use truth as a criterion is determined not only by the world but also by our traditions of inquiring about the world, of assessing the legitimacy of descriptions and claims made about the world, and of evaluating actions performed in the world. So suppose Philomena is reviewing a paper submitted to a scientific journal. By submitting the paper for publication the paper's author has made public the claims that the paper purports to be knowledge. Philomena now becomes engaged in her role as part of the peer-evaluation mechanism by using her knowledge of the field and its norms. Ideally, the journal's editor has chosen her because she is knowledgeable in this field. In her role as public evaluator Philomena will use her knowledge of acceptable practices to critically assess the experimental paradigm the author used to generate the data. Relying on established canons of statistical reliability, she will look at evidential inferences from data to conclusions, If the author's data do not accord with what she knows, she may even attempt to reproduce the experiment. In assessing the author's conclusions she will draw on her knowledge of other writings that she and most others in her field take to be authoritative. And she will use what they say plus her knowledge of the cannons of reasoning to examine the inferences the paper's author makes. In this example and in each of the three aspects of knowledge (acquisition, memory, and use), it is impossible to determine where the social begins and the cognitive ends. The processes by which learning takes place are socio-cognitive; the representation and content of what is learned has many social dimensions (and these to an indeterminable degree); and finally, the actions through which knowledge is exhibited and demonstrated (that is, through which what has been learned is counted as knowledge) rely on public recognition and approbation using socially established cannons of acceptability.

Demonstration

Another dimension of the social aspect of knowledge concerns other people's acceptance and subsequent use of assertions and knowledge claims. Demonstration of knowledge is always publicly manifested, and it leads (in some cases) to public dissemination, such as through published scientific findings. Such public presentation is a necessary condition for others' being able to use a scientist's claims as knowledge and to draw their own inferences based on those claims; that is, an individual's insight is communicated to people in the relevant knowledge community and so made public. The insight is ignored by some and appropriated by others for use in further theorizing.

One important use of knowledge made public is critical interchange or "dialogue," the novel construction of ideas through social interaction, or what has been termed "intellectual interdependency as constructive communication" (Valsiner and van der Veer 2000, 11). Longino (2002) makes a similar point when she writes, "Those assumptions are epistemically acceptable which have survived critical scrutiny in a discursive context characterized by . . . (1) availability of venues for and (2) responsiveness to criticism, (3) public standards (themselves subject to critical interrogation), and tempered equality of intellectual authority" (206). In any of these cases of interpersonal exchange leading to a new mutual understanding of what is being discussed, the extant criteria for what is acceptable play a role along with individuals' reworking of their positions and claims by trying to assimilate or accommodate what has been newly learned. The dynamics of such dialogic or critical construction are complex and variegated, and there is never any assurance of success even when mutual agreement becomes established. Such critical interchanges return us to social aspects of acquisition, for much important learning takes place by critically thinking about and then incorporating what others have said and shown, in science and in all fields. Learning is not confined to conditioning or training of one person or organism by another but often results from cooperative activities that themselves result in mutual restructuring of knowledge representations and structures (as in sense 2, memory, above.)

Such interchanges occur at a social level, too. Sometimes the "institutionalization of a particular knowledge-construction device" (Valsiner and van der Veer 2000, 5) proceeds without significant controversy, as in establishing statistics in social science such that this means of analysis becomes a shared goal and methodological cannon. At other times, instruments and procedures are the subject of heated debate. Scientific controversy may lead to acceptable, profitable, or "progressive" scientific change (see, for example, Machamer, Pera, and Baltas 2000). For example, the matter of properly training psychological laboratory subjects erupted into a "violent controversy" in the 1890s, with functionalists arguing that Wundtian subjects were trained to produce reaction times that were supportive of (required by) Wundt's theory (Boring 1929, 314). In this case the very structure of the data base used as evidence and test of the theory implies isolation of data into parameters that reflect the categorical scheme used by the scientific practitioner. This scheme has a social history and is subject to change. There always must be an argument from data to phenomena to show that, however the data were generated, they somehow are sufficient to capture the phenomenon of interest. Too many experimental techniques have generated great data only to end up on the slag heap of irrelevant science, not because the data are bad, but because the experimental techniques used to get the data are irrelevant to the phenomenon that interests us. Relevant to the Wundt-functionalist controversy is that the ascendance of functionalism and eventually behaviorism in psychology rendered the meticulously collected data of introspectionist procedures extraneous to scientific psychology (see, for example, Watson 1913). Similarly, new paradigms for studying perception made irrelevant the abundant data culled from tachistoscope experiments. At stake is what is to be counted as knowledge.

Some Other Implications

Having briefly mapped out some of the important, nontrivial senses in which scientific knowledge is social, we think we should say a little about why some people consider "the social," at least in the sense of social values, to be something that might contaminate scientific knowledge. Presumably, this potential for contamination is why many people think that separating the cognitive from the social is, in principle, a good thing and keeps science pure.

Longino's analysis emphasizes that *social* seems to be taken as contrasting with *objective*, in the sense that the social is equated with values, that is, *agendas*, that will contaminate and distort the scientific process that searches for objective truth. Values are associated with "wishful thinking" (Longino 2002; Lacey 1999) that promotes scientific production in the service of personal motives: politicized science or science for personal gain. As examples, Longino points to the racist and sexist motivations underlying nineteenth-century agendas in anthropology and biology (and we could include psychology as well).

Yet, although the traditional definition of *objectivity* has been integrally bound to truth, the operationalizing of this concept in science has always had a social or collective dimension. That is, what any group takes to be objective is that on which there has been sufficient intersubjective agreement among observers (see, for example, Kerlinger 1986). Here what is social — that is, what constitutes the intersubjective "checking" of observations and assessing their relations to models or theories — is what is assumed to keep science from slipping into value-laden territory or serving particular agendas. The guiding principle is that as the peer group acceptance of observations and theory-evidence relations increases, the fit of observation and theory with the world may be assumed to increase. A fundamental aspect of scientific practice is inter-rater reliability of measurement. Note also that acceptability (often called "validity") of measurement is similarly socially determined to varying degrees, notably as to what constitutes acceptable criteria, in terms of practices, instruments, and precision, for some measurement (such as of intelligence or action potentials). We shall not discuss here the necessity or efficacy of multiple perspectives, but it is in this context and that of the critical dialogic that such a discussion should occur. There is, then, no contrast between social and objective values that rightly may be drawn here. We could not call objectivity, for example, as it is operationalized, a cognitive value. In fact, it seems more fruitful to say that the social is built into the very conception and application of objectivity, and hence into the scientific ideal and cannons of practice.

One final note will take us back to the algorithmic calculatory rules with which we began. It seems that even here we cannot really rule out the social. Wesley Salmon (1990) and others have suggested that even the theory-evidence relationship is not sacred, since personal and perhaps even social values may enter into these algorithmic rules, such as Bayes's theorem. If such a rule contains a utility function representing personal probabilities or has a variable representing background knowledge or prior probabilities, then it seems most probable that the social features we have spoken about are used to establish those priors or evaluate that background knowledge. If this is true, then we would have the personal or social in the epistemic even in this most restricted place.

REFERENCES

Aristotle, *Nicomachean ethics*. 1998. Trans. D. Ross, Rev. J. L. Ackrill, and J. O. Urmson. Oxford: Oxford University Press.
Boring, E. 1929. *A history of experimental psychology.* New York: The Century Company.
Douglas, Heather. 2000. Inductive risk and values in science. *Philosophy of Science* 67:559–579.
Hempel, C. G. 1965 [1960]. Science and Human Values. Reprinted in *Aspects of Scientific Explanation*.
Kerlinger, F. 1986. *Foundations of behavioral research.* Fort Worth, TX: Holt, Rinehart and Winston
Kuhn, Thomas. 1973/1977. Objectivity, value judgment, and theory choice. In *The Essential Tension*, 320–39. Chicago, IL: University of Chicago Press.
Lacey, Hugh. 1999. *Is science value free? Values and scientific understanding.* London: Routledge.
Laudan, Larry. 1984. *Values and science.* Berkeley: University of California Press.
Longino, Helen E. 2002. *The fate of knowledge.* Princeton, NJ: Princeton University Press.
———. 1990. *Science as social knowledge: Values and objectivity in scientific inquiry.* Princeton, NJ: Princeton University Press.
Machamer, Peter, and Lisa Osbeck. 2003. Scientific normativity as non-epistemic: A hidden Kuhnian legacy. *Social Epistemology* 17(1):3–11.
Machamer, Peter, Marcello Pera, and Aristides Baltas. 2000. *Scientific controversies: Philosophical and historical perspectives.* New York: Oxford University Press.
Salmon, Wesley. 1990. Rationality and objectivity in science, or Tom Kuhn meets Tom Bayes. In *Scientific Theories*, vol. 14 of *Minnesota Studies in the Philosophy of Science,* ed. C. Wade Savage, 175–204. Minneapolis: University of Minnesota Press.
Valsiner, J., and R. van der Veer. 2000. The social mind: Construction of the idea. Cambridge: Cambridge University Press.
Watson, J. 1913. Psychology as the behaviorist views it. *Psychological Review* 20:589–97.
Wittgenstein, L. 1953. *Philosophical investigations.* Translated by G. Anscombe. Oxford: Basil Blackwell.

5

Transcending the Discourse of Social Influences

Barry Barnes
Department of Sociology, University of Exeter

Some Ancient History

In many parts of the sociology and history of science there is now a deep reluctance to refer to "social influences on science," and rightly so. Yet scholars in these fields have long debated the extent and importance of such influences and have viewed the issues involved as straightforward, even if contentious. Historical studies have documented instance after instance wherein "social factors" or "influences" have affected scientific work, whether beneficially or adversely. Instances of adverse influence have included spectacular cases, such as the thalidomide affair, in industrial laboratories, and others, no less striking, involving academic contexts and esoteric topics, like preferred interpretations of quantum mechanics. Choices of research topic, of method, and of theory have all been linked to so-called social factors, and so, too, more significantly as far as philosophers are concerned, have the judgments involved in the weighing of apparently conflicting bodies of data, in interpreting that data, and in deciding whether putative data should be acknowledged as data in the first place. Indeed, variations in what are taken to be the basic criteria of scientific judgment and the fundamental standards of good scientific practice have been related to "influences" operating at the collective as well as the individual level in science, in episodes extending over long periods of time as well as in isolated incidents. Certainly, a philosopher would be likely to find material consistent with the idea that influences have sometimes diverted professional scientists from conformity with whatever basic standards she reckoned to be definitive of properly scientific practice. Thus, if these studies are to be believed, the effect of social influences on science has been profound as well as pervasive, and not merely that of an occasional incentive to the production of disreputable work. Moreover, the work of sociologists could be taken to offer yet further support to this claim, by documenting how scientists from different institutional locations are prone to disagree systematically with one another on technical matters germane to the interests of their employers, and by reminding us that most scientists do indeed now work in contexts where the utilitarian concerns of the employing organization are expected to guide what they do.[1]

Why then should we now be ill at ease when speaking of social influences? It is not that we have come to doubt the relevant historical and empirical findings, but rather that long debate and extended reflection has led to an ever-increasing awareness of the serious problems associated with that form of discourse. The evidence that professional scientists may be strongly (and adversely) affected by the social contexts within which they work can be very impressive. John Braithwaite, for example, in his well documented study of corporate crime in the pharmaceutical industry, noted that, "Data fabrication is so widespread.... as to support an argot—the practice is called 'making' in the Japanese industry, 'graphiting' or 'dry labeling' in the United States" (1984, 57). We have not subsequently come to question whether there was graphiting in the pharmaceutical industry or even whether scientists routinely acknowledged its existence. Rather, we have become wary of speaking of "social influences on science" even in cases like this.

What is the source of this reticence? It lies in the requirements that have to be met if the discourse is to be employed with anything remotely approaching rigor. First and crucially, if "science" is to be "socially influenced," then it must be identifiable as something distinct and separate from the "social" realm, or from the rest of that realm if science is regarded as a part of it. Second, potential "influences" must be located in the realm identified as independent of "science itself." And finally the effects of these influences must, of course, be documented, which ideally requires knowing how "science itself" would have gone in the absence of the influences, under its own dynamic. In addition, if "social" influences are taken to be undue influences, then there is a need to show that when they are operative their consequences are undesirable. Debate and reflection have increasingly led to the conclusion that these requirements cannot be satisfied.[2]

If we look to the distant origins of the relevant debates, we find that social influences were indeed at one time assumed to be undue influences, disturbances to the proper unfolding of science. Science was thought of as an individual activity properly moved only by reason and experience, and "the social" was an improper influence on this individual activity simply by virtue of being social. Matters lie differently today, of course. The assumption that the social is no more than an undesirable external influence on an essentially asocial science has now largely been discredited.[3] Science is now widely recognized as in some sense or other a social activity, although in just what sense remains an open question, and for philosophers especially the conviction that it remains ultimately based on the traditional combination of (an individualistically conceived) reason and experience has not disappeared. Hence, while the contrast of "the social" and "the scientific" continues to be made as if science were an asocial entity, and "social influences" on science do continue to be remarked on, the unspoken assumption is now more often than not that external or extrinsic influences on the internal (social) activities of science are thereby being referred to. The discourse of social influences, in so far as it persists, persists as a part of the familiar internal/external frame: in that frame, a "social" influence is an external one and a source of undue influence by virtue of being external and not by virtue of being "social."

The traditional internal/external frame can incorporate a conception of science as some form of social activity. It may regard it as social activity of some distinct and special kind, and as susceptible to external disturbance. But the problem remains of what makes it special. What characteristics permit it to be demarcated from the rest of the social realm? And debates on this issue have long been riddled with confusion and ambiguity, much of which derives from the clashing thought styles of the various disciplines involved. In keeping with their predominantly naturalistic orientation to the sciences, the studies of historians and sociologists have tended to address science as the institutionalized practice of an occupation or (in earlier times) a recreation, whereas philosophers have often thought in terms of an idealized conception of science. But practice and ideal have not been clearly distinguished, which has inevitably led to different understandings of the role of externalities in science. Philosophers, in particular, have been unduly indulgent in allowing their defenses of science in ideal form to be read as vindications of science as we know it. But historians and sociologists have also contributed to the problem by referring to practices and judgments in many different contexts as "science" without properly reflecting on what has prompted this description.

In addition to the tension between these frames, however, are problems that continued to surface within them as attempts were made to identify science as a distinct, well-defined entity. On the philosophical side of the debate the difficulty was showing why any particular account of science in ideal form should be preferred to possible alternatives — accounts advocating different criteria of good practice and sound judgment, perhaps, or different values to employ in theory choice. And, indeed, there are several competing accounts here, and showing why one should be preferred to another has proven to be difficult. On the empirical side of the debate, the same difficulty was (and is) experienced in another way. When science, or putative science, is studied empirically, it is the activity of human beings that is addressed. Nothing is visible other than people doing things together: communicating with each other; acting, separately and jointly, on the material world around them. Empirical studies claiming specifically to be studies of science must somehow agree on the best way to analyze speech and action into science or scientific research on the one hand and externalities on the other.

In practice, naturalistic studies tended to focus initially on the boundaries and demarcations identified and enforced by those being studied. The distinction between science and externalities drawn by practitioners themselves was the starting point for reflection and analysis. And this starting point typically yielded a picture of scientific fields as bounded domains of practice and culture, set in the larger "external" context of the wider society and its overall culture. But of course "science" as culture and practice may be incompletely differentiated from "other" varieties of these things,[4] and even if this is not so, there may be ambiguity and disagreement about what specific fields should be considered sciences, and what distinct forms of social practice and bodies of knowledge should count as scientific, even among recognized scientists themselves. Indeed, the difficulties encountered by the naturalistic approach to the study of the boundaries of the sciences may be severe, and should not be underestimated. They involve scrutiny not only of established de facto demarcations between the sciences and the "society" outside them but also of the currently fluid and unsettled discourse within which reference is made to a burgeoning "technoscience" (Pickstone 2000) and the emergence of recognized fields of study like biotechnology. Indeed, reflecting on the many and various conceptions of the science-technology relationship, what they have been used to justify and how they have changed over time, is a good way of appreciating how different conceptions and exemplars of science are likely to persist even in what we take to be "science itself."[5]

In summary, the demarcation problem remains unsolved and has all the look of an insoluble problem. And if science cannot satisfactorily be demarcated, then sources of influence independent of it cannot be identified, and their untoward effects on "science itself" cannot be lamented. Already it should be clear why there is so much unease about the discourse of social influences. But, of course, the discourse continues occasionally to be deployed, probably because, whatever its formal failings, it is in tune with what once were widespread informal intuitions. Indeed, it is probably the continuing dualist conviction that "the social" stands in opposition to "the rational" that sustains what remains of the discourse of social influences even as the untenability of that dualist perspective has increasingly been recognized.

In view of this it is worth making brief mention of the particular contribution of sociologists to the relevant debates. From the inception of sociology of science as a specific field of study in the immediate postwar period its practitioners drew attention to the positive role of "the social" in relation to the sciences. A number of studies of the beneficial effects of "external social influences" were initiated by Robert Merton, Joseph Needham, and others at this time. Merton's early work (1970, 1973) was especially interesting and important, in that it linked the "scientific revolution" of the seventeenth century, the mythical point of origin of "modern science," on the one hand, to new demands for particular kinds of technical predictive knowledge, and, on the other, to a particular ethos of social values. (It can go without saying in the present context that the possibility of "external" social values "influencing" science beneficially is one that is still being actively explored.)

Even more salient in the present context was work on the social characteristics of "science itself." In this work the traditional dualist frame came close to being inverted, and "the social" was identified as a bulwark against individual "influences" that would otherwise have been bound to disturb and disrupt the operation of science. Although it was accepted, indeed taken for granted, that the production of scientific knowledge involved individual reason and experience, "the social" was also regarded as intrinsic to the process, something inseparable from what was good in the sciences and actually necessary to guarantee their *objectivity*. And, indeed, there is now a long tradition of work in sociology that has identified the peculiar merits of the natural sciences with how they are ordered socially; and it offers interesting precedents for current philosophical accounts of objectivity, such as Helen Longino's, in chapter 7 (this volume).

The fount and origin of sociological work of this sort is again Merton, who pointed to the existence of an institutionalized normative order in all the sciences that, among other things, enjoined specific forms of disinterest and organized skepticism on individual practitioners. The crucial role of the social and institutional valorization of the methodological standards definitive of science was particularly stressed by Merton, who identified it as a functional necessity in the absence of which individual contributions replete with lapses in reasoning and biased judgments could never be synthesized into a body of objective knowledge. But many variations on this theme were to emerge in the subsequent literature.[6] Thus, the same equation of epistemological virtue with the social and cultural dimension of the sciences is also visible in the work of Thomas Kuhn (1970, 1977), but here a shared, collectively enforced respect, not for supposedly universal standards, but for a specific body of inherited knowledge, is regarded as crucial. If Kuhn is right, then collective acknowledgement of a received set of exemplary achievements or paradigms is the necessary normative basis for the production of objective knowledge, and objectivity involves and is partly constituted by the intersubjective agreement within a specific community on what will count as their exemplars. The intrinsic importance of shared standards and exemplars continues to be emphasized in the sociology of science today, for example, in the work of the so-called Edinburgh school (Barnes, Bloor, Henry 1996). But here respect for standards only as manifest in the practice of peers, and for exemplars only as actually deployed and articulated by them, is identified as the epistemological necessity. And this means that in the last analysis the respect of scientists for *each other* is what makes their research a normatively ordered activity, and that their intersubjective agreement is what permits the objective judgment that an exemplar has been correctly applied (Barnes 2002).[7]

There is no need to dwell on how studies of this sort were initially received. Because they tended to be naturalistic in their approach, suspicious of the idealization and reification of science, and above all, I suspect, because they were expressions of profoundly anti-individualistic modes of thought, many readers were unable to recognize them as genuinely appreciative accounts of the sciences and insistently misunderstood them as attempts to undermine their standing. Kuhn, in particular, long suffered as a result of the widespread prejudice that to describe science as intrinsically social activity could only be to diminish it and to deny the objectivity of the knowledge it engendered. Fortunately, however, this kind of misconception is no longer a major problem. Even though opinions continue to differ on just what makes them so, and indeed on whether or not they are so, it is no longer at all implausible to suggest that the sciences are indeed intrinsically and irreducibly social activities, or even to argue that just this is what makes them admirable activities. That much at least has emerged from the laborious and convoluted debates that used to surround these issues.[8]

Another Perspective on Science and "the Social"

Philosophical styles of argument often involve fastening hard onto some relatively simple and uncontroversial claim or assumption and exploring where a dogged and unrelenting attachment to it may lead. This approach can be very useful in cutting through the accumulated complications surrounding issues like the present one. With luck it may lead to the heart of the matter and circumvent what with hindsight become lesser points or even distractions. I shall adopt this approach here to short-circuit the complexities of the old debates and develop a fundamental criticism of the entire discourse of "social influences on science." Accordingly, I shall begin in the manner of a philosopher, with a simple stipulation: that scientific knowledge is empirical knowledge, applicable to specific empirically identifiable particulars. I expect that this assumption will be uncontroversial, but to take it seriously can very rapidly lead to challenging problems. Indeed, simply to focus attention on the relationship between knowledge and an empirical particular to which it putatively applies is to confront head on one of the most difficult problems of epistemology.

The problem is that relations of sameness between empirical particulars are intransitive: if one thing is the same as another thing, and that thing is the same as a third, it does not follow that the first thing is the same as the third. Unlike a relation of essential identity, which is transitive by fiat, a relationship of mere empirical sameness between two things does not preclude their being different, that is, empirically different, from each other. And the temptation here to speak of the two things as completely the same in some specific empirical feature or property, even if different in others, has got to be resisted, because sameness relationships between the features or properties will themselves be intransitive, and reference to them will merely initiate a regress. The intransitivity of empirical sameness is beyond remedy (Hesse 1974; Kuhn 1977). Thus, when a particular, A, is categorized empirically as of sort X, so that existing knowledge of Xs may be applied to A, the question may arise of whether A is in fact an X after all, that is, whether A is indeed relevantly the same as existing acknowledged instances of Xs, and to that question there is no given and predetermined answer. Because the sameness relation is intransitive, there is no way of decisively refuting the claim that A is not, or may not be, relevantly the same as existing Xs.

It is perhaps fortunate that the intransitivity of empirical sameness is a familiar and uncontroversial matter, since an extended discussion would have been entailed otherwise. Even so, a standard illustration may be useful as a reminder of what is involved. Begin with two tins of paint, one red, one yellow. Make a third tin by mixing equal amounts of the two paints and place it between the first two. Make further tins by mixing each adjacent pair of what is now a row of three, and place them between the tins from which they were made, to make a row of five. Continue in this fashion until any two adjacent tins are indistinguishable by whatever empirical methods or techniques exist. If adjacent tins can be distinguished, continue to mix as before until such time as they cannot be mixed. In terms of empirical sameness, adjacent tins are then just the same; for all existing practical purposes they are identical in color. Yet if the tins are treated as essentially identical in color, so that the color relationship is transitive, one may end up asserting that red paint is identical in color to yellow paint.

Problems associated with intransitivity always arise whenever particulars are classified in terms of their empirical characteristics, and/or empirical knowledge is applied to them: there is not any fact of the matter that indefeasibly determines whether or not a given act of classification is correct and hence whether the associated knowledge does or does not apply in that particular case. But despite being uncontroversial, this is a point that is often hurriedly passed over or obscured. Thus, the application of general knowledge to particular cases is often treated in abstract discussions as a wholly unproblematic and routine matter, a simple deduction from general category to particular case, but of course it always involves far more than this because the question of whether that particular case truly belongs in that general category has to be addressed. In the last analysis this is a question of whether the particular is relevantly the same as those existing particulars to which the general knowledge is already reckoned to apply, a question that the particulars themselves cannot answer because of the transitivity problem.

This question only infrequently becomes a pressing one in actual practice. Most of the time individuals seem to apply their shared knowledge competently as a matter of routine, in sufficient agreement with each other for their sharing of knowledge to continue, save only in a very few so-called borderline cases. Empirical knowledge is applied as though sameness is transitive and trouble fails to arrive. The problems of sameness that could arise in principle apparently fail to arise in practice. And accordingly the transitivity problem is sometimes regarded as of no great importance, one that gives rise only to tiny and trivial differences of judgment on the margins of scientists' practice. This view, however, represents a major misconception, encouraged by an unduly individualistic view of what is involved. A holistic vision is needed here, encompassing the extension of intransitive sameness relations in the practice of entire collectives. Only then is it possible to grasp the profound significance of intransitivity, and what it implies, not only for the notion of "social influence" but also for other topics of central relevance here, including the nature of objectivity, the relationship of "the social" and "the epistemic," and the role of values in science.

If sameness relations are imagined running chainlike from instance to instance, with every link involving analogy not identity, connecting particulars that are different as well as the same, then the differences between instances, even if so very "tiny and trivial" as to be imperceptible, will be liable to accumulate as the chain is extended within the collective, even if the actions involved are entirely routine and a *matter of course*. Individuals would eventually be applying knowledge, Crusoe-like, in radically different ways, if their judgments of sameness were made independently; or, to put the point more precisely, they would eventually lack (shared) knowledge to apply. That each of them was separately proceeding more or less automatically, perhaps without even noticing any difference at all between one instance and the next, or that each was proceeding with extreme care and caution and extending concepts only to what was "obviously" the same as existing recognized cases, would not preclude a journey into solipsism and a disintegration of knowledge into so many individual representations.

In fact, solipsism and the disintegration of (shared) knowledge seem never to have emerged from the routine practice of scientific fields, a situation suggesting that scientists are not independent rational individuals who apply existing knowledge separately, with no more coordination than that which "empirical reality" itself provides. Perhaps the point is an obvious one, but reflection on the transitivity problem conveys it vividly and forcefully, and indicates, moreover, how this coordination is needed everywhere and all the time, not just on special occasions when a choice of method is to be made, say, or a transition of theoretical frame is under way. For a sociologist like myself, the most plausible hypothesis with which to account for this ongoing ubiquitous coordination is that scientists (like all other human beings) are indeed not independent rational individuals, but social agents, whose coherent routine practice is a tribute to their sociability.[9] I want to suggest that the routine use and application of empirical knowledge is a social process in a profound sense, and that the fundamental sociological importance of intransitivity is readily appreciated, perhaps most readily when entirely routine scientific work is considered. Continuing agreement in the routine extension of relations of empirical sameness, such as is necessary where the continued use of shared knowledge is involved, is only possible because the relevant individuals are sensitive to how others are extending those same relations, and ready to realign their own practice in the light of that, and that alone. A continuing mutual susceptibility of this kind is, I suggest, the basis of the essential normative dimension of shared empirical knowledge, which normativity is continually sustained and reconstituted in the interactions of members of the relevant community.

The transitivity problem arises because scientific knowledge must be applicable to empirical particulars, and its discussion focuses attention on the applications of science, or so it may seem. Certainly, it invites reflection on what is conventionally called "applied" work, work that tends often to be neglected in contexts like the present one. The use of animals as surrogate humans in medical research, of artificially produced nucleic acid in efforts to learn about the natural stuff, or of wind tunnels and models to simulate vehicles and aircraft in their normal working environments may come to mind as relevant exemplars. Yet more interesting, perhaps, are problems of what counts as the same as what may be experienced as real and pressing in laboratories devoted to testing and analysis, wherein tissues are matched, DNA "fingerprints" are compared, biopsies are scrutinized for signs of malignancy, viruses or strains of viruses are identified, materials are assayed and identified as more or less pure samples of this or that substance, and so forth. All of these are examples within which transitivity problems obviously lurk, social interaction exists through which a shared practice is sustained notwithstanding, and overt controversy over sameness may nonetheless occasionally erupt. All are examples where insiders sustain a sense of what counts as "the correct application of scientific knowledge" but where outsiders cannot hope to distinguish it from the knowledge itself.

Endless fascinating problems surround science as applied knowledge. For example, just as untried technology notoriously generates problems the first time it is employed "for real," so may empirical knowledge. Eventually, of course, most "routine" applications become smoother and simpler and more recognizably routine as the initial kinks are ironed out. Locally operative inductive feedback loops enable the practice in which the knowledge is applied to settle into what we regard as its optimal form. Those who turn out, ex post facto, to fare worst at identifying malignancy may leave the testing laboratory; those who do best may become the second level evaluators of "problem cases" identified by others. Routine work may become more impressively efficacious as a result of this kind of thing. And the temptation to identify this local achievement of particular people as the correct way of applying the knowledge they started with and its outcome as what the knowledge "really implied" will then be faced and often yielded to.

Fascinating though they are, however, these examples encourage the thought that peculiar difficulties arise specifically at the point where scientific knowledge is applied. And although I can make the point here only as an aside, I want to suggest that this is incorrect. Indeed, I have spoken of knowledge in application only as an expository device. In my view it is close to being redundant to speak of knowledge in this way. It is not just that empirical and historical accounts of the applications of knowledge, even accounts of routine laboratory testing, assaying, analyzing, and so forth, are also accounts of the growth of the knowledge itself, and that applications are the sources of the data used to validate what is being applied. More profoundly, it is that knowledge may not even be identifiable as something independent of its applications: certainly, the most penetrating general accounts of the nature of scientific research refuse to make a strong distinction between the one and the other.

Consider Kuhn's (1970, 1977) general account of scientific research, according to which exemplars, and not theories and bodies of data, are the fundamental elements in which scientific knowledge consists and the elementary units in terms of which it is transmitted.[10] Research has to proceed from instance to instance, in moves whereby solved problems with exemplary status provide the models with which to address as yet unsolved ones. A scientist will typically seek to solve a problem by setting it into analogy with an already solved problem, an exemplar. Existing knowledge will be applied in the form of an exemplar, and any prima facie solution thereby produced will be more widely evaluated by reference to other exemplars. Analogies between the supposed solution and the repertoire of existing ones, as perceived by other practitioners, will be the normative reference points when its standing as knowledge is determined, so that what appertains in practice will also appertain in epistemology; and knowledge application and knowledge evaluation will not be identifiable as discrete and separate processes. Crucially, on this account, scientific research has to proceed by the application of knowledge constituted as (exemplary) applications. And of course it is no surprise that someone with this view of research should have manifested a lifelong curiosity about the social relations of scientists and the nature of the communities that they constituted. Once it is recognized that knowledge consists in existing applications with the status of exemplars, that research is the further application of existing exemplars, and that such exemplars permit themselves to be applied further in different ways, one is easily led to the thought that authority located in the collective itself may be the crucial element in specifying what will count not only as exemplars but also as the correct ways of further applying and extending them.

In summary, the problem posed by the intransitivity of sameness pervades scientific practice. And although opinions may differ on where precisely the problem is encountered and on just why it is important, that it is both irremediable and of fundamental significance is hard to deny. Moreover, the problem is clearly not limited to the application of concepts and empirical generalizations. Laws, rules, norms, standards, theories, models, exemplars, and, particularly salient here perhaps, *values* all give rise to versions of the problem in their application to particular cases.[11] None of these applies itself nor comes with an owner's manual offering instructions that fix decisively its correct use in the next case. Formally speaking, human beings retain discretion over how to apply any of these to the next case, that is, over what will count as a correct application in that case. Loosely speaking, the speech act involved may be thought of as making an analogy between the next case and those that have gone before, and there is no indefeasibly correct way of making an analogy.

However one chooses to formulate the point here, it is clear that its consequences are far-reaching. Intransitivity exposes a profound open-endedness in the use and application of knowledge. It is like the Duhem-Quine problem of indeterminacy reconstituted at the micro-level, the level below that of theories and, indeed, below that of concepts. And as such it not only creates problems when we seek to justify the knowledge of a scientific field but also actually puts fundamental difficulties in the way of specifying what the knowledge implies as well as what the knowledge is that the field supposedly shares. Thus, it is perhaps unsurprising that so many philosophical accounts pass over the problem at high speed. As a former chemist I know something of the strength of the desire to look away from mere sameness relationships, to the identity relationships between essences with which the field of chemistry is replete. Relationships of identity, and the social processes of reification that engender and sustain them, are themselves of great sociological and philosophical interest: in collectively constructing identity relationships as conventions, often by converting strategically important sameness relationships into them, scientists help to engender the coherence necessary to the collective practice of their field. But it is an unduly reified vision of science that is produced when only identity relationships are considered, and for all that reification is itself of enormous interest and in need of empirical study as a social process, too much of sociological relevance is obscured if only the products of that process are considered and sameness relationships are ignored.[12]

To focus attention on the intransitivity of sameness and the consequent open-endedness in the use and application of knowledge makes the manifest orderliness of the practice of the sciences into a challenging sociological problem. Once it is accepted that knowledge cannot itself convey "the" correct way in which to apply it in any specific instance, even to the most rational and cognitively competent individual user, the problem of why users apply it as they do, and how they generally manage to agree in how they apply it, so that "it" continues to exist as shared knowledge, becomes one of understanding the contingent actions of human beings. All the problems involved in identifying and understanding the normative dimension of knowledge become associated with every particular action to which it is correctly applied. Every such action is made visible as a social action in a profound sense. And the knowledge that informs it ceases to have a history independent of its carriers and users. Indeed, to understand the normativity that makes knowledge knowledge, as opposed to mere individual belief or intuition, one must constantly and continually look to those carriers and their interactions.

Reflections on the Argument

Recall that one of my purposes in raising the problem of the intransitivity of sameness relationships was to cut through to the core of the problem of "social influences on science." I hope that even those who disagree with my conclusions will recognize that I have done this. Because of intransitivity, there is no specific path along which the practice of science would proceed in the absence of "influence," no trajectory that can be inferred from existing knowledge or existing methodological rules. Similarly, there is nothing that can indefeasibly be demarcated as science and no realm indefeasibly identifiable as external or extrinsic to it wherein influences on it might be located. There are only human beings, seeking to put a shared cultural inheritance to use, whether to directly facilitate the prediction and control of features of the external world or to test and further refine the inheritance itself. In light of this we should cease to employ the discourse of social influences; nothing of what is required for its reputable employment actually appertains.

It might be objected that, even if valid, all this amounts to a pedantic and unforgiving criticism of what can be a useful and readily understood form of speech. Is it really harmful, in practice, to speak of graphiting as the result of social influences on science? Is it wrong to ask how extensively scientists are affected by factors defined as illegitimate or irrelevant within their particular professional cultures, and how frequently their work is corrupted, or else inspired, as a consequence? The answer here must be, first of all, that there is neither need nor justification, even in the most informal discourse, to the term *social* as a near-synonym for *improper, undue,* or *illegitimate* — or even for *external, extrinsic,* or *nonscientific*. Second, if *social* is nonetheless insistently misused in this way, then the result can only be to misdirect our understanding and impoverish our imagination: "the social" is not an optional extra as far as scientific practice is concerned, still less a distorting intrusion. And, finally, if we wish to know how far the activities of professional scientists are affected, not by "social factors" but also by factors of these latter kinds, then the answer to this empirical question is easily given: professional scientists are extensively and sometimes profoundly affected by factors describable in these terms, or so the evidence suggests.[13]

Needless to say, the case against referring to "social influences on science" can only be stronger still where formal, academic discourse is concerned. Certainly, historians and social scientists should resist all temptation to employ such discourse and recognize as well the serious limitations of the larger internal/external frame into which it has been incorporated. They do, of course, need to take account of "internal/external," and even "scientific/social" as actors' categories, used by professional scientists to establish the boundaries and distinctions necessary to the standing of their fields as subcultures carrying valued and credible expertise. But these categories should not be part of their own framework of analysis. As such they would merely subvert the tasks of naturalistic description and explanation appropriate to these fields. Sociologists and historians should be content simply to document the sequences of contingent actions that constitute scientific research, and perhaps to attempt to explain them and their trajectory naturalistically, that is, in terms of causal antecedents.[14] A naturalistic approach of this kind transcends the misleading discourse of social influences altogether; and insofar as the approach aspires to be explanatory, it seeks to provide causal accounts of research as a social process.[15]

In truth, however, recommending these conclusions to historians and sociologists is scarcely appropriate, since a monistic and naturalistic approach is and has long been very widely employed in these fields. As is usual, practice has preceded and inspired the theoretical discussions within which it is rationalized and justified. The best history of science, in particular, has long provided examples of the kind of approach I have tried to describe.[16] So I shall move on and consider the philosophical significance of what has been said, something that can scarcely be avoided in the present context even by a nonphilosopher like myself. No doubt, much that is relevant to historians and sociologists, with their naturalistic concerns, will be irrelevant to those philosophers and epistemologists who conceive of their disciplinary goals as normative ones. Certainly, both the knowledge claims of scientists and their conduct of research may be addressed by philosophers without attention being paid to their causal antecedents; and there is no clear reason why a discussion of social influences should necessarily interest them. Even so, there may be reasons why this specific discussion will be of interest, given its claim that a normative dimension is inherent in the activity of applying knowledge as well as that of evaluating it, and its insistence that scientific research is intrinsically a social activity.

What I am claiming here does indeed clash quite strongly with what is presumed in a great deal of philosophical work, despite the fact that the intransitivity of sameness, from which much of my argument derives, is a familiar theme in philosophy. Philosophers tend to be especially concerned with how scientists evaluate knowledge, whether to identify what accounts for the "success" of science or, more rarely, to reappraise those evaluations as a part of an independent epistemological project. But the focus of this concern is generally extremely limited, and it can sometimes even be asked how far it is directed to things identifiable with the actual practice of science at all. Those who seek the secret of the success of science still largely concentrate on the testing of theories by reference to a given body of evidence, and on the criteria and standards relevant to identifying the best theories in light of this evidence. Such an approach chooses to pass over, for whatever reason, the normative dimension of the processes where evidence is collectively constituted as evidence, and the no less important normative dimension of the processes where what are accounted the applications and implications of theories come to be accepted as such applications and implications. Most, some would say all, of the social activity that scientific research actually consists in disappears from view in consequence. And the conjecture that the success of science may primarily be related to the specific form of its social organization and its position as specialized goal-oriented activity in a larger institutional order is deprived of its most plausible sources of support in advance of inquiry.[17]

Philosophical efforts to pinpoint the secret of the "success" of science have been impoverished by a failure to treat scientific judgments as contingent social actions. But both these efforts, and efforts to critically evaluate science in specifically "philosophical" terms, have also been impoverished by a failure to treat philosophical judgments in the same way, perhaps because to do this greatly intensifies the problems faced by philosophers in justifying those evaluations. Philosophers have always faced the problem of showing why the standards, values, and criteria of evaluation they employ should be preferred to different ones; but now they also need to show why their applications of those standards should be preferred to different ways of applying them, and to recognize that appeal to the standards "themselves" can be of no help in doing this. Moreover, the predicament they face here is generally more serious, at least in a practical sense, than the parallel predicament scientists face. Those scientists who participate in the practice of a specific field, and have inherited its cultural tradition, will be familiar with exemplars of any standards, values, and criteria they reckon to share and, through established social relations with each other as competent peers, will usually interact in ways that sustain coordination and agreement in practice as analogies with these exemplars are made. Thus, they will be in a position to apply standards, values, and so forth, just as they apply empirical knowledge, extending (intransitive) sameness relationships to new instances just as they would do with concepts or laws. As a social network able to agree on how to proceed in analogy with exemplary cases, they can point to *that* agreement as representing what the standard or value or criterion "itself" implies in the next instance. But philosophers who would evaluate the practice of a scientific field in terms of "their own" standards, or whatever, usually face far greater problems in establishing agreed conclusions and securing standing for them.

The naturalistic approach of sociologists is notoriously difficult to reconcile with that of philosophers, and many readers of this analysis will accordingly find it unsurprising that its conclusions are difficult to reconcile with the assumptions and aspirations of at least some of the most important extant approaches in the philosophy of science. Even so, I do not myself believe that there need be any tension between the two fields, and I would not wish what I have said to be misinterpreted as in any sense an expression of skepticism about the salience and importance of work in philosophy of science. Indeed, I prefer to see whatever value there may be in my analysis as testimony to the possibility of fruitful links between sociology and philosophy. After all, my argument centers on the problem of the intransitivity of sameness, and the most profound and penetrating discussions of that problem have been provided precisely by philosophers of science (Goodman 1977; Hesse 1974; Quine 1960). Whatever kind of admirable philosophical project it has been that has taken that problem as intrinsically interesting, and produced such invaluable reflections on it, dovetails beautifully with current projects in the sociology of science.

NOTES

1. For older case studies, see Forman (1971) and Farley and Geisen (1974). Nelkin (1992) provides both relevant case studies and valuable general discussion. Older critical discussions of "influences" and the associated internal/external frame include those of Barnes (1974, chap. 5) and Shapin (1982); see also Schuster (1990) and Porter (1990). A lack of current literature explicitly reflecting on influences and the internal/external frame reflects the way that the relevant discourse has lost respectability and been, as my title puts it, transcended, at least as a resource for serious naturalistic studies of science.

2. There is also the interesting terminological point that graphiting, along with all other forms of "bad science" or "junk science" or "unduly influenced science," is not science at all, and should be described in something other than this oxymoronic vocabulary. From here, it can be argued as well that historical studies of "influence" tell us little or nothing about how science, involving properly scientific judgments, is "influenced" by "social" factors, since by definition "science" excludes anything so influenced. In practice, however, awareness of this point seems not to have discouraged use of the offending vocabulary.

3. When I originally read this sentence while presenting a version of this chapter in Pittsburgh, one commentator responded that here was a truth so obvious as to need no mention; another commentator suggested my claim was close to the opposite of what actually appertains empirically. No doubt my view of the matter, like that of others, has something to do with my institutional location and disciplinary affiliation.

4. So, for example, when the work of Thomas Malthus (the "political theorist" and "ideologue") was identified as an external "social influence" on Charles Darwin's "science," a helpful influence as it happens, a part of the subsequent debate concerned whether anything like a clear cultural or social divide separated Darwin and his work from Malthus and his. And since "science" was not then identified as a specialized

occupation, or even differentiated as a set of practices, the problems raised about the "influence" of Malthus were difficult and fascinating. See Young (1969).

5. There is no need for those involved in empirical studies to be concerned about this. It is not their task to resolve any disagreements they encounter, or to clarify any gray areas, concerning boundaries. They may confine themselves to documenting an incomplete consensus, taking note of whatever contestable and ill-defined boundaries are established around and between specific scientific fields in actual practice, and monitoring the traffic to and fro across those boundaries, such as they are. For this approach in practice, see Gieryn (1999).

6. For all that their approach is profoundly different, recent fascinating historical studies by Shapin and Schaffer (1985) and Shapin (1994) also give expression to this Mertonian theme. See also Pinch (1990).

7. Much of the constructivism currently predominant in the sociology of science may also be placed in this same tradition, since it identifies the achievements of the sciences as collective accomplishments.

8. Much more has resulted in some contexts. In the sociology of science in Europe, for example, talk of "social influences" or even "external" influences has all but disappeared. Some of the arguments that led to this disappearance are closely paralleled with those to be advanced in the next section here, although the authorities cited tend to be different. Garfinkel (1967) and Wittgenstein (1968) are key figures in this context; see also Bloor (1997). The great popularity of the insistently monistic philosophy/sociology of Bruno Latour (1987, 1988, 1993) has also been significant in the turn away from what essentially was a dualist form of discourse.

9. I have discussed the large issues involved here as entirely general sociological problems in Barnes (1995, 2000).

10. There are other interpretations of Kuhn of course. Barnes (2002) sets out the interpretation alluded to here, and something close to it is set out in the philosophical accounts of Sneed (1971) and Stegmuller (1976). See also Giere (1988).

11. Habermas puts the point nicely in relation to norms: "No norm contains within itself the rules for its application. Yet moral justifications are pointless unless the decontextualization of the general norms used in justification is compensated for in the process of application. Like any moral theory, discourse ethics cannot evade the difficult problem of whether the application of rules to particular cases necessitates a separate and distinct faculty of prudence or judgment" (1990, 206). Unfortunately, however, he cannot bring himself to acknowledge the profundity of the problem and still insists on treating norms as analogous to the propositions that constitute, in his view, truth claims in science. Propositions are statements with determinate meanings, meanings independent of how they are used or who believes in them. If norms are propositions, then what particulars they correctly apply to may be inferred from the norms themselves, without the mediation of context, and the sameness relation is transitive. But if "no norm contains within itself the rules for its application," as Habermas insists, then propositions-represent illegitimate reifications of verbal activities, and the sameness relation is intransitive, which he refuses to accept. Habermas's use of a reifying propositional frame

for his moral philosophy, even though he is very well aware of the problems it faces, echoes the responses of strong rationalists to the same problems in the philosophy of science.

12. Lest this be seen as a swipe at realism per se, I should add that reification is often a good thing, indeed that it is an essential thing: the emphasis in the main text should be placed on the "unduly."

13. See note 1.

14. Complete accounts of this kind are not to be looked for, of course. How scientists act will presumably be a consequence of their own individual powers and propensities, their interaction with the material world, and their social interaction with other human beings, especially (given that shared scientific knowledge is being spoken of) with *each other*. But knowledge of actions and their antecedents is knowledge of particulars, and the arguments from intransitivity already deployed cycle back and apply to it. Even so, partial and selective accounts of scientists' actions as ordinary empirical phenomena, affected, say, by background and training, by social affiliations, by interactions in the work place, by the idiosyncratic features of the instruments and techniques they employ, and so forth, may still be worth having, and may help to account, for example, for why some scientists act differently from others in applying what ostensibly is the same knowledge.

15. Note that links and connections constituting the social process of research itself are also to be conceptualized causally: this, of course, is why the internal/external distinction is of no basic importance and why even its pragmatic use can so easily lead us astray.

16. Among innumerable examples are Crosby Smith (1998), Rudwick (1985), and Olby (1994). *The path to the double helix* is indeed worthy of particular mention. This account of how different inheritances of knowledge and competence were used and extended can be read almost as an account of a continuing struggle with intransitivity. And within it "the social" is nowhere and yet everywhere. Those who peruse it will quickly find that there is no way of separating "the scientific" from "the social" in the story it tells.

17. I have no wish to advocate this hypothesis, although it has been important in the context of the sociology of science. It needs to be borne in mind that rather than identifying science and noting its success we often identify success and call what is successful science, opening up the possibility that science is a potpourri of valued activity possessed of no single "secret."

REFERENCES

Barnes, B. 1974. *Scientific knowledge and sociological theory*. London: Routledge.
———. 1995 *The elements of social theory*. Princeton, NJ: Princeton University Press.
———. 2002. Thomas Kuhn and the problem of social order in science. In *Thomas Kuhn*, ed. T. Nichols, 122–41. Cambridge: Cambridge University Press.
———. 2000. *Understanding agency: Social theory and responsible action*. London: Sage.

Barnes, B., D. Bloor, and J. Henry, eds. 1996. *Scientific knowledge: A sociological analysis*. Chicago: Chicago University Press.
Bloor, D. 1997. *Wittgenstein, rules, and institutions*. London: Routledge.
Braithwaite, J. 1984. *Corporate crime in the pharmaceutical industry*. London: Routledge.
Farley, J., and G. L. Geison. 1974. Science, politics, and spontaneous generation in nineteenth-century France: The Pasteur-Pouchet debate. *Bulletin of the History of Medicine* 48:161–98.
Forman, P. 1971. Weimar culture, causality, and quantum theory, 1918–1927: Adaptation by German physicists and mathematicians to a hostile intellectual environment. *Historical Studies in the Physical Sciences* 3:1–115.
Garfinkel, H. 1967. *Studies in Ethnomethodology*. Englewood Cliffs, NJ: Prentice Hall.
Giere, R. N. 1988. *Explaining science*. Chicago, IL: University of Chicago Press.
Gieryn, T. 1999. *Cultural boundaries of science: Credibility on the line*. Chicago, IL: University of Chicago Press.
Goodman, N. 1977. *The structure of appearance*. Boston, MA: Reidel.
Habermas, J. 1990 *Moral consciousness and communicative action*. Cambridge: Polity
Hesse, M. B. 1974. *The structure of scientific inference*. London: Macmillan.
Kuhn, T. S. 1970. *The structure of scientific revolutions*, 2nd ed. Chicago, IL: University of Chicago Press.
———. 1977. Second thoughts on paradigms. In *The essential tension: Studies in scientific tradition and change*, 293–319. Chicago, IL: University of Chicago Press.
Latour, B. 1987. *Science in action*. Cambridge, MA: Harvard University Press.
———. 1988. *The pasteurization of France*. Cambridge, MA: Harvard University Press.
———. 1993 *We have never been modern*. Sussex: Harvester.
Merton, R. K. 1970. *Science, technology, and society in seventeenth-century England*. New York: Harper.
———. 1973. *The sociology of science*. Chicago, IL: University of Chicago Press.
Nelkin, D. ed. 1992. *Controversy: Politics of technical decisions*. London: Sage.
Nickles, T. ed. 2002. *Thomas Kuhn*. Cambridge: Cambridge University Press.
Olby, R. 1994. *The path to the double helix*. New York: Dover.
Olby, R. C., G. N. Cantor, J. R. Christie, and M. J. S. Hodge, eds. 1990. *Companion to the history of modern science*. London: Routledge.
Pinch, T. 1990 The sociology of the scientific community. In *Companion to the history of modern science*, ed. R. C. Olby, G. N. Cantor, J. R. Christie, and M. J. S. Hodge. London: Routledge.
Porter, R. 1990. The history of science and the history of society. In *Companion to the history of modern science*, ed. R. C. Olby, G. N. Cantor, J. R. Christie, and M. J. S. Hodge. London: Routledge.
Quine, W. V. O. 1960. *Word and object*. New York: Wiley.
Rudwick, M. J. S. 1985. *The great Devonian controversy*. Chicago, IL: University of Chicago Press.

Schuster, J. A. 1990. The scientific revolution. In *Companion to the history of modern science*, ed. R. C. Olby, G. N. Cantor, J. R. Christie, and M. J. S. Hodge. London: Routledge.

Shapin, S. 1982. History of science and its sociological reconstructions. *History of Science* 20:157–211.

———. 1994. *A social history of truth*. Chicago, IL: Chicago University Press.

Shapin, S., and S. Schaffer, 1985. *Leviathan and the air-pump: Hobbes, Boyle, and the experimental life*. Princeton, NJ: Princeton University Press.

Sneed J. 1971 *The logical structure of mathematical physics*. Boston, MA: Reidel.

Smith, C. 1998. *The science of energy*. Chicago, IL: University of Chicago Press.

Stegmuller, W. 1976. *The structure and dynamics of theories*. New York: Springer-Verlag.

Wittgenstein, L. 1968. *Philosophical investigations*, 3rd. ed. Oxford: Basil Blackwell.

6

Between Science and Values

Peter Weingart
Institute for Science and Technology Studies, Bielefeld University

(Re-)Defining the Problem

Addressing the issue of science and values inadvertently leads to a host of well-worn themes, most prominent among them the discussion between Popperian philosophy of science and the sociology of knowledge over the impact of social values and interests on scientific knowledge.[1] That discussion has lasted for more than half a century without coming to a satisfactory solution; this suggests that something may be wrong with the definition of the problem. Niklas Luhmann (1990) has advanced a radical break with this line of debate by accepting the circularity of knowledge as constitutive of any knowledge production (294). In this view, the claim of the sociologists of (scientific) knowledge that values and interests have an impact on scientific knowledge is trivial. After all, science is a social enterprise. How could it be otherwise? But the conclusion that sociologists of knowledge, and proponents of laboratory studies in particular, draw from their case material, namely that there is no "interesting epistemological difference" between the procedures of science and other institutional areas, is blatantly wrong (Knorr-Cetina 1992, 408). Although it may be difficult to define in unambiguous terms what exactly separates scientific from other types of knowledge, the institutional evidence is clear. The public character of scientific knowledge, of scientific method, and of scientific discourse in general, as well as the fact that there is no fixed barrier stopping critical examination, have all proved to be most successful. Indeed, this is so much the case that these characteristics of scientific knowledge, first institutionalized in a separate functional subsystem devoted to the pursuit of "truth," have subsequently begun to permeate the rest of society, shaping communication patterns in other subsystems as well. Of course, we know of all kinds of limitations and abuses of the open communication of scientific knowledge (some of which I will discuss later), but it does not make much sense to deny the existence of the institutional *difference* between scientific and other types of knowledge as a social fact. Thus, the issue of science and values in its old form may be cast aside and put to rest.

Why, then, and in what sense does the relationship between scientific knowledge and values remain problematic? Among the several answers that may be given, I will focus on one: The interesting issues are emerging on the *institutional level*, that is, at the institutional interface between science, on the one hand, and politics and the media, on the other. Politics and science are interconnected in countless arrangements based on the assumption and expectation that decisions can and should be grounded on allegedly secure and objective knowledge. Actual practice reveals something very different. Knowledge in its instrumental and in its legitimating functions for decisions of any kind proves to be neither secure nor objective or unequivocal. The closer the interface, the more science becomes politicized. The last three decades have seen a continuous loss of authority of science as a source of reliable knowledge. At the same time the reliance on science and the expectation of obtaining reliable knowledge from science have grown. How can this paradox be explained? What, if anything, is new about the arrangement between science and politics that accounts for the loss of authority?

In the following I will first conceive of scientific knowledge as being relative to different meanings and what I call "contexts of relevance." Then I will address procedural changes in the relationship between science and its social environment that can account for the increased politicization. Thus, I take the interface between science and politics as a strategic research site for the exploration of the relationship between scientific knowledge and social values.

The Relationship between Science and Values in a Context of Relevance

A crucial question at the outset is where scientific knowledge and values interact. To answer this question, it is necessary to distinguish between two types of communication: that which is involved in producing scientific knowledge and that which is involved in implementing it in the broadest sense.

On the basis of this distinction, a starting point is the seemingly trivial insight that scientific knowledge and its purported intersubjectivity rests not on the individual scientist's good intentions and abilities but is the outcome of a social mechanism, that is, open communication, in which every opinion is subject to criticism (Popper 1962, chap. 22) In Popper's (and Merton's) versions, the notion of the hypothetical nature of scientific assertions and the institutional mechanism of the public discourse of science that is open to any criticism at all times is obviously an ideal, typical construct (Merton 1957). It evidently has influenced Habermas's as well as Luhmann's radical move to conceptualize "society" as communication, thereby shifting the object of study altogether. The public nature of the scientific discourse that Popper calls the "social aspect of scientific method" may be questioned on the basis of empirical observation. What are some factors that intervene in the process?

First, almost too obvious to be worth mentioning is the structured nature of the scientific discourse. The scientific discourse is never completely open, and therefore knowledge is always "revealed" to some extent, to use Popper's term, that is, it is always a smaller group of individuals who have made certain discoveries and advance certain truth claims rather than the relevant community of experts as a whole. The range from "revealed" to "public" knowledge, or from individual revelation to the public discourse, is at best a continuum. What starts out as the discovery of an individual or a small group of researchers may or may not become certified knowledge of the entire community. That process involves time, and, above all, the time required to criticize, test, and accept or reject the initial truth claim. In the interim the status of that claim is tenuous; it is subject to judgments that are heavily influenced by social rather than intellectual criteria: status of authority, prestige of institution, proximity of field, personal acquaintance, and so on. This is most obvious and well researched in the case of "peer review," that is, the very process of certifying knowledge by open criticism from competent others (USGAO 1994). In this sense one may speak of the *social structure* of the communication process not being significantly different from that of the rest of society, at least not in principle. The crucial difference is, however, that the selective operation of these structural criteria does not have any legitimacy. On the contrary, while in society at large they are contested, it is considered illegitimate to appeal explicitly to any of them in the process of scholarly criticism. This demarcates the communication in science from that in other realms of society. (It is a very worthwhile question, one that cannot be pursued further here, to consider the extent to which the values that dominate the scholarly communication process have permeated society at large).

Second, the communication of truth claims that would ideally be insulated from any outside interference until closure is reached and the truth claim is established within the scholarly community is not isolated in that way in actual fact. In the interval between the advance of a truth claim and its acceptance, the knowledge in question permeates into the public discourse and becomes part of the socio-cultural frameworks that guide social action. In particular, it constitutes specific expectations and choices among the general public.

The recent discourse on reproductive technologies, molecular medicine, and, lately, promising stem cell research are cases in point. I will later come to the institutional mechanisms underlying this phenomenon, but for the moment I focus on the impact that the promises and perspectives of stem cell research and other related technologies have on ethical values, as well as the possibilities they create for — and the decisions they force on — individuals and entire societies. When embryos suddenly assume potential value as a source of cell tissue for developing cures to genetic diseases, the choice to use them for that purpose is, first of all, constituted by that knowledge. A choice is made, even if the opportunity is rejected. Also, the previously existing value system does not remain unaffected, if only because a new, ethically ambivalent choice has been added.

The process of scientific criticism, although not defunct, has a much reduced influence on what is being done with and to the knowledge concerned. How immediate this transfer of knowledge into the public discourse is depends on a combination of two factors: the dominating issues of that discourse at any given time and the content of the knowledge produced. For example, the latest advances in classifying comets may go unnoticed while a premature release about a statistical correlation between blood groups and behavior patterns makes headlines in the popular press and is shortly thereafter translated into the recruitment policy of industrial firms. Public attention may be of no significance to the scientific communication process, yet it may affect it. In the case of the public debate over the permissibility of stem cell research in Germany, the respective scientists and their funding bodies became deeply involved in the discussion that ended with a decision by the federal parliament laying out a legal framework for the regulation of that research. Of course, such a framework falls short of intervening directly into questions of true and false but only pertains to the ethical and medical desirability of that research and the definition of the corresponding legal constraints.

Third, scientific knowledge is used in political decision-making processes of all kinds and here assumes a dual role. It is *instrumental* in solving specific problems, and it *legitimates* certain positions in political debates and/or contested decisions that have been made or are to be made. The protracted public debate over the safety and the economics of nuclear energy in the 1970s demonstrated to the public for the first time that scientific experts could be recruited for particular positions and interests entertained by governments, industry, and opposing citizens' movements. Also, the public became witness to specialized academic debates designed to legitimate highly politicized decisions. After the 1986 accident at the Chernobyl nuclear power plant in the Ukraine, local communities in Germany that had acquired Geiger counters to measure radioactive fallout were prohibited from publishing the information they obtained and eventually had to hand over their instruments. This was possible but not mandatory under regulatory law governing the nuclear industry, and it demonstrated the government's enormous fear of the consequences of factual knowledge getting in the hands of the general public. There are other pertinent examples: An extreme case of instrumentalization of scientific knowledge for legitimating purposes can be seen in U.S. industry's multimillion dollar attempt to recruit experts for the purpose of undermining the dominant opinion in the scientific community that human-generated carbon dioxide emissions stemming from burning fossil fuels cause global warming.[2]

These and innumerable other examples attest to the fact that scientific knowledge carries great prestige and is both welcomed and feared as support or de-legitimation of political interests and social values. In modern democracies political decisions have to be legitimated in two ways: delegation of power by popular vote and rationality of action in light of scientific knowledge (Roqueplo 1995; Ezrahi 1990). Politicians who act contrary to available knowledge will have a difficult time maintaining their legitimacy, as became apparent in the case of South African president Mbeki's contesting in public the established thesis that HIV causes AIDS and, as a consequence, denying millions of patients available medication.

This case is an interesting though admittedly rare illustration of the clash between the political and the scientific modes of communication. Central to the HIV/AIDS debate is the pattern of attack and defense of the boundary between science and politics. The debate was sparked by Mbeki's decision to set up an international AIDS panel to advise his government. Several prominent "dissidents" were members of the panel. In justifying its decision in response to protests from the medical community, the government argued explicitly with reference to "organized skepticism" that is, the central norm of science that the scientific discourse be kept open to all dissenting voices until the process comes to a standstill, when no new ideas or evidence are put forth. The invitation of dissenting voices seemed to be in line with this norm. The scientific community, on the other hand, appeared to break this rule by trying to silence the dissidents and to close the debate. Thus, the appeal to the dissidents by Mbeki and his government, as well as the attacks on the scientific establishment, all appear to be in defense of scientific freedom. It is as if the respective roles of politics and science are inverted, politicians defending the freedom of scientific discourse and scientists attempting to suppress dissenting voices. However, this is only the superficial appearance.

In actual fact, the scientific community considers debates closed when the consensus appears strong enough and no new claims are considered relevant. Attempts to intervene are then considered an illegitimate disturbance, especially if supported from outside. In this view, Mbeki illegitimately intervened in the process when he called in dissidents to re-open the case of HIV/AIDS, as they would not have had any attention, let alone a public forum, without his support. Closing a controversy in science is important in order to progress, but it also creates a state that is very vulnerable to interference from outside, as the achieved consensus may collapse at any time in response to new evidence. Thus, the norm of keeping the debate open to all voices only applies as long as the discourse is under way. Once it appears sufficiently closed, the achieved consensus is defended against any attempt to destabilize it again, because this is a threat to established knowledge. We do not know whether Mbeki did not realize this, or whether he consciously applied the norm in order to take advantage of the dissidents' prestige as scientists, however limited, to legitimate his own position. As the debate continued, pressure on Mbeki mounted, and it became evident that his intervention into the "scientific process" threatened to develop into a crisis of political legitimacy for himself and his government.

Fourth, some scientific knowledge is continuously being implemented before "all the facts are in" and/or the assumed certainty of knowledge proves to be premature. In essence this means that uncertainty of knowledge is endemic with respect to knowledge that is ultimately applied. In the latter case the lack of knowledge is not apparent while the implementation takes place leading to unforeseen risks and judgments after the fact. Loren Graham has justly observed that "in chronological terms the most dangerous period of the development of a science is when enough is known to advance the first fruitful speculations and to try a few interventions, but not enough is known to bring discipline to these speculations or to predict the possible side effects or after effects of speculation," and he cites the example of the early days of human genetics, when "pseudo-scientific eugenic theories and practices" were based on the infant science (Graham 1981, 379).

Given its constitutive uncertainty, putting scientific knowledge to practical use implies risks. When society as a whole is subjected to such risks or this happens continuously on a sufficiently large scale, one can speak of "society as a laboratory" (Krohn and Weyer 1989). In fact, the social distribution of knowledge production, that is, the generalization of the reflexive generation and use of knowledge as a mode of action beyond the narrow realm of the scientific laboratory or the introduction of some risky technology, may be said to be the defining criterion of the "knowledge society." It is characterized by the growth of knowledge production and a new dynamics of innovation triggered by it. However, contrary to general expectation this does not lead to ever-growing certainty and rationality, but rather to a new perception of risks, uncertainty, and lack of knowledge. The awareness of contingency and of the apparent paradox that more knowledge generates more lack of knowledge has superceded the naive belief in the connection between growth of knowledge and certainty of decisions.

This heightened awareness of contingency created by science is now affecting science itself. It reveals the problem areas for which no knowledge exists or where the existing knowledge is insufficient and insecure. Not only have naive expectations of an ultimately rational state of society and the rule of science been put to rest, but also evaluation, interpretation, and judgment have been reinstated as inexorable aspects of any decision making. An obvious indicator of this changed perception of scientific knowledge and its lack of ultimate certitude is the increasingly pervasive discourse on the need for more practical knowledge, a greater proximity to the "real" problems of the world through inter- or transdisciplinarity, a better accessibility of knowledge and the call for its "social robustness."

Elsewhere I have called the above mentioned four types of interchange between scientific knowledge and different types of social application or impact the "context of relevance" in order to juxtapose it to Popper's "contexts" (Weingart 1988). What is described with the "context of relevance" attests to the ubiquitous utility of scientific knowledge and its diffusion into virtually all realms of life. The risks associated with science, the lack of knowledge with respect to certain contexts, are the result of this success, that is, the application of scientific knowledge or scientific methods to problems that are actually too complex. The capabilities of science are overstretched, expectations about its potential too high. Recent debates about "post-normal" science, or "Mode 2," reflect this when they proclaim a new kind of science governed by ethical principles, that is, values (Funtowicz and Ravetz 1993, 121; Gibbons et al. 1994).

The significance of this will become apparent when we look at the institutional change that has accompanied the emergence of this "new order of knowledge." The gist is that application and implementation of knowledge are, more than ever before, governed by the reflexive awareness of the risks involved and of the lack of knowledge for preempting them.

Institutional Correspondences of the "Context of Relevance"

What I have described so far as the "context of relevance" was deliberately focused on the nature of scientific knowledge and its relationships to "communicative environments," that is, focused on the fact that scientific knowledge is asymmetrically distributed, that it constitutes choices and may have immediate repercussions on established values, that it may legitimate or delegitimate decisions, and that it may orient action, although being premature. These are properties that are specific to knowledge with a high truth value and indicate interfaces between that knowledge and its environments. In the remaining section I want to point to some institutional changes that are responsible for the widespread impression that scientific knowledge has become diluted by "values" or is finally ruled by them, whatever one's ideological preference.

The first such change, underlying the others to be mentioned, is the shift of Western industrial societies from bourgeois to mass democratic societies. This process, initiated by World War I and given impetus by World War II, has, first of all, led to a generalization of participatory claims. These go hand in hand with an enormous expansion of education and with a replacement of birth as a criterion of social status with educational achievement. Status mobility and a constantly shifting social structure are as much an expression of this development as are new forms of participation in the political process through public movements.

Another consequence is the equally impressive expansion of the media during the second half of the twentieth century. Not only has television been added to the print media and become the most popular medium, but also the media industry has expanded in size, serving a larger and more diverse public. This means that societies are observing and describing themselves with a much higher intensity than they were fifty years ago; consequently, they communicate at a much higher rate than they once did. This is just the backdrop for some other developments.

A major institutional change caused by mass democratization has taken place in the realm of politics. It pertains to the use of scientific knowledge in political decision making, that is, the establishment of expert advisory bodies in government administrations. Political controversies are now being carried out by recourse to expert knowledge. Access to this expert knowledge is no longer the privilege of governments but is now also shared by opposition parties as well as a diversity of nongovernmental organizations engaged in all kinds of issues. As this happens, political contenders in any controversy with a scientific-technical base (this includes, of course, medical and social scientific knowledge as well) have been driven to compete for the latest knowledge. However, the paradoxical outcome is that, in principle, the competition for the latest, and therefore supposedly most compelling, scientific knowledge drives the recruitment of expertise beyond the realm of consensus knowledge right up to the research frontier where knowledge claims are uncertain, contested, and open to challenge. Thus, the scientification of politics and of regulatory and legal practice literally produces its opposite, the politicization of the expertise involved.

The pervasive discourse on risk and uncertainty of knowledge must be understood in light of this mechanism. The closer the integration of scientific knowledge into policy making, the higher the expectations of certainty, reliability, and safety. At the same time, the political reliance on that knowledge systematically exceeds its potential, thereby pushing it beyond the limits of certainty. The *loss of distance* between science and politics characterizing this new arrangement is best exemplified by the fact that many technological developments (nuclear energy, genetically altered plants and food, waste disposal) are implemented experimentally. The hitherto institutionally separate process of experimentation in the laboratory had to achieve closure before implementation would begin. In many cases, due to the complexity of technology involved, this separation is now suspended, and the respective technologies are implemented with considerable uncertainty—expressed in the probability of failure—as to their actual performance.

In spite of the fact that the paradoxical outcome of the competition for scientific expertise, that is, the production of uncertainty, is widely noted, the arrangement as such and the correlating concepts of science, politics, and the advisory process remain surprisingly stable. The reason for this is the lack of an alternative: There is no other system of more reliable knowledge and consequently none providing higher legitimacy for political decisions. But two responses can be observed on the side of institutional politics and science. One response may be called *contraction*. It is the attempt to limit the access to scientific expertise and/or to "re-privatize" (that is, close off from the public) controversies until closure is achieved. This is, in effect, an attempt to *regain* the *distance* that has been lost to democratization. Since, once established, democracy cannot easily be repealed, this distance has to be regained within its context, rendering chances of success very small.

Another strategy may be called *expansion*. Although it appears to be contradicting the first, it is actually complementary. It, too, is a reaction to the abundance of scientific expertise in policy making. One variant is the deferral of controversial issues to open and staged discourses, be it in the media, in "roundtables," in mediation processes, in committees of all kinds inside and outside the parliamentary system. By involving other types of expertise, be it bearers of "local knowledge" or value-oriented voices such as ethicists, in a mass democratic context the legitimacy of the process is enhanced. Discursive forms of dealing with public controversies are considered an advance over technocratic forms when an "aloof elite" of scientific experts makes pronouncements that are not subject to criticism. This has to be understood in light of the fact that in modern societies risks are increasingly attributed to the decisions of people, rather than seen as "dangers" posed by nature (on this essential distinction, compare Luhmann 1991). This means that the orientation to the future has shifted from "dangers" to the perception of "risks," and the perceptions are being given voice by those who are affected by decisions that constitute risks. Because of the diversification of the observation of risk, that is, the possible consequences of decisions, scientific expertise loses its privileged position and monopoly of definition. The discursive arrangements, in particular, give the impression that "values" have entered the advisory process and are an inextricable part of scientific knowledge.

Just as in its relationship to politics, science's relationship to the media is experiencing rapid and fundamental change. Here, too, we can observe close coupling between science and the media. According to the dominant hierarchical concept of types of knowledge, scientific knowledge is superior to popular, everyday knowledge. Science has a monopoly on truth; the media merely transmit information. In this arrangement, the control over the adequacy of this transmission lies with science. To most scientists, popularized knowledge is in the best case simplification, in the worst case pollution (Hilgartner 1990, 519ff; Green 1985). This concept of popularization implies a passive and generally unspecified public, and the process of communication from science to the public is conceived of as being unidirectional. The public is excluded from the production and validation of knowledge and by implication is considered to be incompetent to judge the transferred knowledge (Whitley 1985, 4). The popularizers, whether authors or the modern media, are not granted an independent function in this asymmetrical model.

In media and communication studies, and more recently in sociology of the media and of science, it has been recognized that this concept of the media as strictly dependent and reflecting a mirror image of reality is naive. Instead, it is recognized that the media operate under their own parameters (profitability, public attention) and, not surprisingly, have developed their own particular strategies of representing the world. Their parameters of operation and their selection criteria have an impact on the way in which news is selected and processed. In this sense it may be said that the media construct their own reality in the same way science does. They just use different instruments, different approaches to "reality," and different forms of representation. Thus, the complaint (especially frequent among scientists) that the media provide "wrong" or "distorted" reports is futile.

To the degree that the media gain more and more importance in shaping public opinion and even public perception of reality, the monopoly of science in judging representational adequacy may be weakened. Science's abstract criterion of truth is now being confronted by the media's criterion of public attention and acclaim. The reliability of information (as represented by the prestige of a scientific journal) now competes with popularity ratings (represented by the circulation figure of a newspaper or the number of viewers of a TV broadcast). The validation criteria of science are not necessarily replaced but are supplemented. Evidence could be provided to demonstrate that the media have gone even further: with the self-assurance supported by the command of mass attention, they posit their own criterion of "emotional appeal" against that of science, "truth" (Weingart and Pansegrau 1999, 9–10).

That scientists have recourse to a nonscientific public is not a novel phenomenon. In a sense, it is part and parcel of popularization. Originally unspecified and diffuse publics were envisioned by popularizing authors. But as time went on these nonscientific publics became more focused and their involvement assumed a more strategic function (Bucchi 1998). They serve the purpose of mobilizing legitimacy with reference to two types of problems: the securing and expansion of the boundaries of science vis-à-vis its social environment and the settling of scientific conflicts in cases where they cannot be solved internally, by scientists. As long as these mobilization efforts are brief and diverse, the narrowing of the distance between science and the media poses little risk for science's self image. It is for good reasons—namely to protect both science's epistemological distinctness and its institutional independence — that part of the unwritten ethos of science is the disdain for gaining prominence by public acclaim through, for example, media stardom or pre-publication in nonreviewed journals, in contrast to reputation building by peer review. Both in terms of sanctioned behavior and of communication, this exemplifies the distance between science and the media. If, however, the sciences' strategic use of the media becomes commonplace, there will in all probability be repercussions for science. The phenomenon at issue may be termed the *science-media-coupling* or the *media-ization* of science. It is typical for modern mass democracies and takes on different forms that I will not pursue here further.

Conclusions

I have dealt with the issue of science and values in a different way than philosophers or historians of science may have expected. In the first part I tried to identify those aspects of scientific knowledge that seem to challenge notions of its being isolated from social values. I want to stress again that this is not an epistemological argument. Rather, the focus is on "communicative relations" of that knowledge, that is, its social environment. One could also refer to Luhmann's argument that knowledge that is not communicated in effect does not exist. When it is communicated, it is inevitably linked to other contexts of meaning and thereby to values, interests, preferences, decisions, and so on. This is an abstract property of knowledge. The criteria I have tried to identify for scientific knowledge may not be exhaustive, but hopefully they delineate the difference between scientific and other types of knowledge.

In the second part, I examined two particular institutional realms, politics and the media, with the purpose of looking at the changes that have taken place in their relationship to science in the context of mass democratization. My conclusion is that one can observe close couplings in each case that explain the loss of the privileged status of science and scientific knowledge during the last half century. This development can also be interpreted as leading into an increasingly important role for social values in science, and it is interpreted in this way by scholars who proclaim the end of "Mode 1" and the new era of "postnormal" science. As a sociologist (of science) I claim two things, however. First, that this development is not interpreted in a fruitful way if one speaks of a "blurring of the boundaries" between science and other parts of society. Both analytically and historically the difference is important in order to understand present developments. Second, in light of my analysis the crucial issue is how, when, where, and to whom scientific knowledge is communicated. In other words, the issue is the relationship between scientific knowledge and its environments. The problem of values in science is thereby recast in a way that makes it accessible to complementary research by philosophers and sociologists alike.

NOTES

1. For the purpose of this analysis, I do not distinguish between values and interests, as I take values to be "preferences" of any kind.
2. *New York Times*, April 28, 1998.

REFERENCES

Bucchi, M. 1998. *Science and the media: Alternative routes in science communication*. London: Routledge.
Ezrahi, Y. 1990. *The descent of Daedalus*. Cambridge, MA: Harvard University Press.
Funtowicz, S. O., and R. Ravetz, 1993. The emergence of post-normal science. In *Science, politics, and morality: Scientific uncertainty and decision making*, ed. R. von Schomberg, 85–123. Dordrecht: Kluwer.
Gibbons, M., H. Nowotny, C. Limoges, M. Trow, S. Schwartzman, and P. Scott. 1994. *The new production of knowledge*. London: Sage.
Graham, L. 1981. *Between science and values*. New York: Columbia University Press.
Green, J. 1985. Media sensationalisation and science: The case of the criminal chromosome. In *Expository science: Forms and functions of popularisation*, vol. 9 of *Sociology of the sciences yearbook*, ed. T. Shinn and R. Whitley, 139–62. Dordrecht: Kluwer.
Hilgartner, S. 1990. The dominant view of popularization: Conceptual problems, political uses. *Social Studies of Science* 20:519–39.
Knorr-Cetina, K. 1992. Zur Unterkomplexität der Differenzierungstheorie. Empirische Anfragen an die Systemtheorie. *Zeitschrift für Soziologie* 21:406–19.
Krohn, W., and J. Weyer. 1989. Gesellschaft als Labor. Die Erzeugung sozialer Risiken durch experimentelle Forschung. *Soziale Welt* 40:349–73.
Luhmann, N. 1991. Verständigung über Risiken und Gefahren. *Die politische Meinung* 36:86–95.
———. 1990. *Die Wissenschaft der Gesellschaft*. Frankfurt: Suhrkamp.

Merton, R. K. 1957. Science and democratic social structure. In *Social structure and social theory,* ed. R. K. Merton, 550–61. Glencoe: Free Press. (Published originally in 1942 as "Science and technology in a democratic order," *Journal of Legal and Political Sociology* 1:115–26.)

Popper, K. 1962. *The open society and its enemies.* 4th ed. London: Routledge and Kegan Paul.

Roqueplo, P. 1995. Scientific expertise among political powers, administrations, and public opinion. *Science and Public Policy* 22, 3:175–87.

USGAO (United States General Accounting Office). 1994. *Peer review: Reforms needed to ensure fairness in federal agency grant selection.* Washington, DC: GAO.

Weingart, P. 1988. Close encounters of the third kind: Science and the context of relevance. *Poetics Today* 9 (1):43–60.

Weingart, P., and P. Pansegrau, 1999. Reputation in science and prominence in the media: The Goldhagen debate. *Public Understanding of Science* 8:1–16.

Whitley, R. 1985. Knowledge producers and knowledge acquirers: Popularisation as a relation between scientific fields and their publics. In *Expository science: Forms and functions of popularisation,* vol. 9 of *Sociology of the sciences yearbook,* ed. T. Shinn and R. Whitley, 3–30. Dordrecht: Kluwer.

7

How Values Can Be Good for Science

Helen E. Longino
Department of Philosophy, University of Minnesota

The Ideal of Value Freedom

Values are good for science — the values of truth, objectivity, accuracy, and honesty in results are integral to most notions of good science. But these are not the values causing concern. The ones being questioned are those that might interfere with the realization of the "good" values: social values and to some extent pragmatic values, ideas about social relations or about social utility that may, without vigilance, be expressed in scientific reasoning or representations of the natural world. While much argument about the role of these sorts of values presupposes that they should not play a role in science, certain epistemological analyses of scientific judgment challenge this assumption. Like many thinkers, I want to urge that philosophy of science include attention to the roles that values, interests, and relationships in the social and cultural context of science play in scientific judgment as well as attention to the impacts of science and science-based technologies on society. I am of the view, however, that in principle there is no way of guaranteeing the eliminability of such values from science. We should stop asking whether social values play a role in science and instead ask which values and whose values play a role and how. But here I want to show not so much how as why values can be good for science.

The recent so-called science wars made this task more difficult by exaggerating the differences between approaches to the sciences that are social and cultural and those that are philosophical. Social and cultural approaches insist on the embeddedness of scientific inquiry in its social contexts and the impossibility of understanding the direction and outcomes of scientific investigation without taking those contexts into account. They argued on the basis of empirical studies of the progress of work in laboratories and research programs that no principled distinction could be made between cognitive and noncognitive elements among the causal factors in scientific judgment (see Bloor 1991; Knorr-Cetina 1983; Latour 1987). Many philosophers of science found themselves in the position of defending the rationality of science against its perceived detractors (see Laudan 1984; Goldman 1995; Kitcher 1993). And so it went on: reason rules versus unreason rules.

Much philosophical discussion about the relationship of science to its social contexts is pursued under the rubric of values in science. This functions as a kind of catch-all for the messy and complex world of social relationships as it might bear on the practice of science. Treating the social dimensions of science as a set of questions about values in science has distracted philosophers from investigating the social dimensions of science. But there is a reason why philosophers worry about claims that seem to undermine the value-neutrality of science. It is worth pausing, therefore, to note why value-freedom has been thought to be an ideal of and for the sciences.

Being free of values is a virtue for science because we want our acceptance of theories to be impartial and not a matter of wishful thinking. In a culture where so much rests on the sciences we fear that certain kinds of values will lead to acceptance of representations of the natural and social worlds in theories, hypotheses, and models that favor the interests of certain members of or groups in society over those of others. The ideal of value freedom is also bound up with the ideal of universality: what counts as a scientific truth or scientifically supported claim for one person or community should count as such for any other, no matter how different their cultural values.

The natural sciences were thought to exemplify the ideal of value freedom because they prescribed or were thought to prescribe methods of hypothesis and theory testing that guaranteed reliance on logic and observation alone, that is, on universal capacities that could be exercised in a content-neutral way. Scientific inquiry pursued rigorously could lead us to accept representations of natural (and perhaps social) phenomena and processes that were free of the taint of metaphysics as well as social biases such as the racism and sexism that infected much of nineteenth-century biology and anthropology.

Equating value freedom with methodological rigor cuts in several ways. Taking value freedom as an ideal led some of us, feminists, antiracists, socialists, to question whether certain scientific research programs were actually value free. Science should be value free, but it is not. Greater vigilance about biases will correct this defect (see Hubbard 1979; Gould 1986). But the value-free ideal has another face: if impartially pursued, value-blind, scientific inquiry produces results that do end up favoring certain groups in society, or that when applied have certain consequences, we must accept those outcomes if they are the result of impartial methods impartially applied. One can see this consequence articulated in the response of advocates of research programs criticized for sexism (Witelson 1985). If science tells us that women are biologically less well equipped than men to do math, well, that's unfortunate, but so be it. This kind of attitude is among the factors that stimulated feminist philosophers to investigate the grounds for claiming that science at its best or in its nature is value free.[1]

Rationality, Sociality, Plurality

The field opened up by the feminist interventions and extended by social studies of science has become crowded and in recent years an unbridgeable rift grew between those who maintain the value freedom of science and those who reject it even as an ideal. Each side of the rift emphasizes its preferred analytic tools to the exclusion of those of the other. As a consequence, accounts intended to explicate the normative dimensions of epistemological concepts, that is, elaborating the relationship of knowledge to such concepts as truth and falsity, opinion, reason, and justification, have been too idealized to gain purchase in actual science, whereas accounts detailing actual episodes of scientific inquiry suggested that either our ordinary normative concepts have no relevance to science or that science fails the tests of good epistemic practice. This cannot be right. The stalemate between the two sides is produced by both sides' accepting a dichotomous understanding of the cognitive and the social.

According to the dichotomous understanding of these notions, if an epistemic practice is cognitively rational, then it cannot be social. Conversely, if an epistemic practice is social, then it cannot be cognitively rational.[2] What further is meant by "rational" or "cognitive," on the one hand, and by "social," on the other, varies from scholar to scholar.[3] The dichotomy between them, however, structures the thinking of a number of writers on scientific knowledge. Elsewhere I tease apart the dichotomy's components and offer an account of epistemological concepts that integrates the rational and the social (Longino 2002a). Among the components of the dichotomy are two contrasting assumptions, monism and nonmonism, about the content of scientific knowledge. I understand monism as follows:

For any natural process there is one (and only one) correct account (model, theory) of the process. All correct accounts of natural processes can form part of a single consistent and comprehensive account of the natural world.

Nonmonism is often treated as antirealism of some kind, but there can be eliminativist, constructivist, and realist versions of nonmonism. This means that there can be two forms of realism: a monist realism, which holds that there is or will be one correct and comprehensive account of the natural world, and pluralist realism, which I understand as follows:

For any natural process, there can be more than one correct account (model, theory) of the process. This is especially likely in the case of complex processes. It is not necessary that all correct accounts of natural processes form part of a single consistent account of the natural world. Rather than one complete account, multiple approaches may yield partial and nonreconcilable accounts.

Philosophers of science who advocate pluralism disagree about the grounds for the view and about the precise nature of the pluralist claim. (For different articulations, see Dupre 1993; Ereshevsky 1998; Mitchell 2002; Rosenberg 1994; Waters 1991.) Those advocating strong forms of pluralism are claiming that the complexity of natural processes eludes complete representation by any single theoretical or investigative approach available to human cognizers.[4] Any given approach will be partial; and completeness, if achieved at all, will be achieved not by a single integrated theory, but by a plurality of approaches that are partially overlapping, partially autonomous, and that resist unification. For example, organismic development can be investigated in different ways that preclude alternative understandings. Insight into the genetic contributions to development is achieved by holding environmental conditions constant. But then one gets no understanding of environmental or other nongenetic factors in development. And vice versa.

Many philosophical accounts of scientific knowledge are incompatible with such pluralism. They assume as a condition of adequacy of criteria of knowledge that there is one uniquely correct account of the phenomenon to be known. Conversely, a standard criticism of pluralism is that it makes knowledge impossible. I contend that accounts of knowledge should not presuppose either monism or pluralism. Whether the world is such as to be describable by one model or many is not a priori decidable. Therefore, one of the constraints on the analysis of knowledge ought to be that neither metaphysical position is presupposed. What would such an account look like?

Knowledge as Social

I propose the social account of knowledge as one way to satisfy the constraint. To see how it does so it is useful to start with the central problem to which that account is addressed: the underdetermination problem. At the heart of philosophical reflection about scientific knowledge is the gap between what is presented to us, whether in the kitchen and garden or in the laboratory, and the processes that we suppose produce the world as we experience it, between our data and the theories, models, and hypotheses developed to explain the data. As long as the content of theoretical statements is not represented as generalizations of data or the content of observational statements is not identified with theoretical claims, then there is a gap between hypotheses and data, and the choice of hypothesis is not fully determined by the data. Nor do hypotheses specify the data that will confirm them. Data alone are consistent with different and conflicting hypotheses and require supplementation.

Philosophers have had a variety of ways of describing and responding to this situation. Pierre Duhem (1954), the first philosopher of science to raise the underdetermination problem (as different from the problem of induction), emphasized assumptions about instruments, for example, that a microscope has a given power of resolution, or that a telescope is transmitting light from the heavens and not producing images internally or not systematically distorting the light it receives. But the content of background assumptions also includes substantive (empirical or metaphysical) claims that link the events observed as data with postulated processes and structures. For example, that two kinds of event are systematically correlated is evidence that they have a common cause or that one causes the other in light of some highly general, even metaphysical, assumptions about causality. The correlation of one particular kind of event, such as exposure to or secretion of a particular hormone, with another, such as a physiological or behavioral phenomenon, is evidence that the hormone *causes* the phenomenon in light of an assumption that hormone secretions have a causal or regulative status in the processes in which they are found, rather than being epiphenomenal to or effects of those processes. Such an assumption has both empirical and metaphysical dimensions. Assumptions of this kind establish the evidential relevance of data to hypotheses. Among other things, they provide a model of the domain being investigated that permits particular investigations to proceed.

Some philosophers discuss underdetermination as a problem of the existence of empirically equivalent but inconsistent theories. The underdetermination under consideration here concerns the semantic gap between hypotheses and data that precludes the establishing of formal relations of derivability without employing additional assumptions. In this picture, different explanatory hypotheses may not have exactly the same empirical consequences, but instead may have some overlapping and some nonoverlapping consequences. If they are empirically adequate to the same degree, then empirical evidence alone cannot serve as grounds for choosing between them. Particle and wave theories of light stand in such a relationship to each other, but there are many other examples as well. The additional (background) assumptions required to establish a connection between hypotheses and data (reports), then, include substantive and methodological hypotheses that, from one point of view, form the framework within which inquiry is pursued, and from another, structure the domain about which inquiry is pursued. These hypotheses are most often not articulated but presumed by the scientists relying on them. They facilitate the reasoning between what is known and what is hypothesized.

I take the general lesson of underdetermination to be that any empirical reasoning takes place against a background of assumptions that are neither self-evident nor logically true.[5] Such assumptions, or auxiliary hypotheses, are the vehicles by which social values can enter into scientific judgment. If, in principle, there is no way to mechanically eliminate background assumptions, then there is no way to mechanically eliminate social values and interests from such judgment. Some sociologists of science used versions of the underdetermination problem to argue that epistemological concerns with truth and good reasons are irrelevant to the understanding of scientific inquiry and judgment (Barnes and Bloor 1982; Pickering 1984; Shapin 1994; Collins and Pinch 1993; Knorr-Cetina 1983; Latour 1987, 1993). The point, however, should not be that observation and logic as classically understood are irrelevant, but that they are insufficient. The sociologists' empirical investigations show that they are explanatorily insufficient. The philosophers' underdetermination argument shows that they are epistemically insufficient.

My view is that rather than spelling doom for the epistemological concerns of the philosopher, the logical problem of underdetermination, taken together with the sociologists' studies of laboratory and research practices, changes the ground on which philosophical concerns operate. This new ground or problem situation is constituted by treating agents/subjects of knowledge as located in particular and complex interrelationships and by acknowledging that purely logical constraints cannot compel them to accept a particular theory. That network of relationships — with other individuals, social systems, natural objects, and natural processes — is not an obstacle to knowledge but a rich pool of resources — constraints and incentives — to help close the gap left by logic. The philosophical concern with justification is not irrelevant, but it must be somewhat reconfigured to be made relevant to scientific inquiry. The reconfiguration I advocate involves treating justification not just as a matter of relations between sentences, statements, or the beliefs and perceptions of an individual, but as a matter of relationships within and between communities of inquirers.

This expansion of justification sees it as consisting not just in testing hypotheses against data, but also in subjecting hypotheses, data, reasoning, and background assumptions to criticism from a variety of perspectives. Establishing what the data are, what counts as acceptable reasoning, which assumptions are legitimate, and which are not become in this view a matter of social, discursive interactions as much as of interaction with the material world. Since assumptions are, by their nature, usually not explicit but taken-for-granted ways of thinking, the function of critical interaction is to make them visible, as well as to examine their metaphysical, empirical, and normative implications.

The point is not that sociality provides guarantees of the sort that formal connections were thought to provide in older conceptions of confirmation, but that cognitive practices have social dimensions. Acknowledging this social dimension has two consequences. In the first place, any normative rules or conditions for scientific inquiry must include conditions applying to social interactions in addition to conditions applying to observation and reasoning. A full account of justification or objectivity must spell out conditions that a community must meet for its discursive interactions to constitute effective criticism. I have proposed that establishing or designating appropriate venues for criticism, uptake of criticism (that is, response and change), public standards that regulate discursive interaction, and what I now call tempered equality of intellectual authority are conditions that make effective or transformative criticism possible (Longino 2002a, 128–35). The public standards include aims and goals of research, background assumptions, methodological stipulations, and ethical guidelines. Such standards regulate critical interaction in the sense of its serving to delimit what will count as legitimate criticism. Thus, these standards are invoked in different forms of critical discussion, but most importantly they are themselves subject to critical scrutiny. Their status as regulative principles in some community depends on their continuing to serve the cognitive aims of that community. The conditions of transformative criticism may not be the conditions ultimately settled on, but I contend that something like them must be added to the set of methodological norms.

Second, even though a community may operate with effective structures that block the spread of idiosyncratic assumptions, those assumptions that are shared by all members of a community will not only be shielded from criticism but also, because they persist in the face of effective structures, may even be reinforced. One obvious solution is to require interaction across communities, or at least to require openness to criticism both from within and from outside the community. Here, of course, availability is a strong constraint. Other communities that might be able to demonstrate the non-self-evidence of shared assumptions or to provide new critical perspectives may be too distant, spatially or temporally, for contact. Background assumptions then are only provisionally legitimated; no matter how thorough their scrutiny given the critical resources available at any given time, it is possible that scrutiny at a later time will prompt reassessment and rejection. Such reassessment may be the consequence not only of interaction with new communities but also of changes in standards within a community. These observations suggest a distinction between a narrow and a broad sense of justification. Justification in a narrow sense would consist in survival of critical scrutiny relative to all perspectives available within the community, whereas in a broad or inclusive sense justification would consist in survival of critical scrutiny relative to all perspectives inside and outside the community.[6]

Clearly, that a theory is acceptable in C at t, in the narrow sense of justification, does not imply that it will be acceptable to C at t_2, or that it must be acceptable to any other community. Furthermore, there is no requirement that members of C reject background assumptions simply because they are shown to be contingent or lacking firm support. Unless background assumptions are shown to be in conflict with agreed on data or with values, goals, or other assumptions of C, there is no obligation to abandon them—only to acknowledge their contingency and thus to withdraw excessive confidence. Background assumptions are, along with values and aims of inquiry, the public standards that regulate the discursive and material interactions of a community. The point here is that they are both provisional and subordinated to the overall goal of inquiry for a community. Truth simpliciter cannot be such a goal, since it is not sufficient to direct inquiry. Rather, communities seek particular kinds of truths. (They seek representations, explanations, technological recipes, and so on. Researchers in biological communities seek truths about the development of individual organisms, about the history of lineages, about the physiological functioning of organisms, about the mechanics of parts of organisms, about molecular interactions. Research in other areas is similarly organized around specific questions.) Which kinds of truths are sought in any particular research project is determined by the kinds of questions researchers are asking and the purposes for which they ask them, that is, the uses to which the answers will be put. Truth is not opposed to social values, indeed, it *is* a social value, but its regulatory function is directed/mediated by other social values operative in the research context.

The possibility of pluralism is a consequence of the possibility of alternative epistemological frameworks consisting of rules of data collection (including standards of relevance and precision), inference principles, and epistemic or cognitive values. Other philosophers have advanced pluralism as a view about the world, that is, as the consequence of a natural complexity so deep that no single theory or model can fully capture all the causal interactions involved in any given process. While this may be the case, the epistemological position I am advocating is merely open to pluralism in that it does not presuppose monism. It can be appropriate to speak of knowledge even when there are ways of knowing a phenomenon that cannot be simultaneously embraced. Whether or not it is appropriate in any given case depends on satisfaction of the social conditions of knowledge mentioned earlier. When these are satisfied, reliance on any particular set of assumptions must be defended in relation to the cognitive aims of the research. These are not just a matter of the individual motivations of the researchers but of the goals and interests of the communities that support and sustain the research. On the social view all of these must be publicly sustained through survival of critical scrutiny. Thus, social values come to play an ineliminable role in certain contexts of scientific judgment.

Values in Science, Again

I maintain that this is an account of scientific knowledge and inquiry (or the basics of one) that both integrates the rational and the social and avoids begging the question for or against pluralism. As to the first, the philosopher is right to see the sciences as a locus of cognitive rationality; the sociologist or sociologically sensitive historian is right to see the sciences as a locus of social interactions (that are not containable within the lab or research site). The mistake is to accept a conceptual framework within which these perspectives exclude each other. With respect to the second, we can talk about knowledge of a phenomenon X made possible by one set of methodological commitments and standards guided by a particular question and also about different knowledge of the same phenomenon made possible by a different set. As long as two (or more) incompatible models of X are working in the ways we want (are narrowly or even broadly justified), why not accept that they are latching on to real causal processes in the world, even if these cannot be reconciled into one account or model? Only a prior commitment to monism precludes this, but whether we end up at that mythical end of inquiry with one true account for each domain or more than one is a matter of how the world is and is neither presupposed nor settled by epistemological reflection.

I now draw some lessons concerning the relationship of science and values. The possibility of pluralism that is part of this account has implications for the ideals both of universality and of impartiality. Universality does not make sense as an ideal except in a very restricted way — results hold for those sharing an investigative framework, cognitive aims, and the values in relation to which a given cognitive aim makes sense. What might be genuinely universal is the judgment that within a framework organized by a particular cognitive goal a given result holds, but this, of course, leaves room for a different result emanating from inquiry differently organized. What about impartiality? One of the aims of many philosophers of science has been, as I mentioned at the beginning, to show how, in spite of the de facto presence of social (and personal) values and interests, scientific inquiry can nevertheless be cleansed of them. The very possibility of pluralism turns the value-free ideal upside down — values and interests must be addressed not by elimination or purification strategies, but by more and different values. To see this, consider the following.

First, suppose pluralism is right, that is, the world is not such as to be in the end describable by one theory or conjunction of theories. Then, even if a given theory has impeccable evidentiary support, is justified in the narrow sense, that it has problematic or noxious social consequences (that is, its acceptance would advance or undermine the interests of one or more groups in society relative to others) is reason not directly to reject it but instead to develop an alternative approach that has equivalent empirical validity. (This is not an armchair pursuit; it takes time, effort, and resources.) The social payoff is an escape route from natural inevitability arguments. The epistemic payoff is an increase in the range of phenomena that we can know or explain. This multiplicationist strategy is constrained, but not foreclosed, by requirements of empirical adequacy. It does not encourage one thousand flowers to bloom, but only two or three may be needed.

So, even if the arguments attributing racial differences in IQ tests were impeccable by the standards of behavior genetics, a commitment to equality or to one's race or sex is reason to explore an alternative explanation, such as Claude Steele's theory of exacerbated performance anxiety or some other. Similarly, feminists' objections to gene-centric or master molecule accounts of biological processes are expressed not just by rejecting them as determinist or reductionist but also by developing alternative accounts. Pluralism affirms the partiality, that is, incompleteness, and not the falsity of gene-centric accounts.

Suppose, on the other hand, that monism is right, that the world *is* describable by one theory or conjunction of uniquely domain-specific theories. Even if this is so, there is no reason to believe it unless those theories that belong in the set have been tested against all possible alternatives, so that a theory's having noxious consequences is again good reason for one with different values to develop an alternative approach. This will increase the alternatives in play and increase the likelihood that eventually, in the long term, we will exhaust all possible alternatives and settle on the conjunction of uniquely correct domain-specific theories.

Feminist interventions in physical anthropology and primate ethology since the 1970s constitute a recent classic example of value-driven research that has improved quality of science in those areas. Feminists have brought new phenomena and data to the attention of their disciplines and have drawn new and different connections between phenomena that were already known to their communities.

The standoff among different research approaches in bio-behavioral sciences offers another example. I have been studying contrasting and competing approaches to the study of human behavior. In particular, I have been looking at differences (and similarities) among behavior genetics approaches and approaches that emphasize aspects of the social environment as explanatory of behavioral profiles or dispositions (see Longino 2001, 2003, and forthcoming). Classical behavior geneticists look for and develop methods for identifying and interpreting intergenerational behavioral correlations ("concordances"), and molecular behavior geneticists look for and develop methods for identifying correlations between genetic structures and behaviors. Social environmental approaches, including family systems and developmental systems approaches, look for and develop methods for identifying environmental and social determinants of behavioral differences. Members of each side characterize the other as politically and ideologically motivated. The social environmentalists accuse behavior geneticists of being socially insensitive, rigidly reductionistic, and giving support to racism, sexism, and social policies that perpetuate racial and gender injustice. Behavior geneticists accuse social-environmentalists of being fuzzy-headed liberals who want to engage in dangerous social engineering. This mutual caricature under- and probably misstates the values involved. In addition to whatever political values are involved, the research is driven by divergent professional interests both within the research communities and in the clientele they serve, by aesthetic values, and by social values and overall conceptions of human nature.

Close examination shows that these approaches are in a narrow way incommensurable. Each parses the space of possible causes differently, so it is not meaningful to compare how well they fit the data. The data and the contexts in which they emerge as data are different from approach to approach. But each program is capable of revealing empirical regularities that the other cannot. The different values that partially sustain each approach ensure their persistence. The consequent plurality of nonreconcilable accounts of the behaviors studied enhances our scientific understanding rather than diminishing it. Human behavior may be so complex that no single research approach can provide complete understanding. Divergent values prevent foreclosure and drive an expansion of knowledge and understanding rather than narrowing them. Persuasive arguments for plurality may also lead policy makers to turn to science for narrowly conceived purposes, but not for general accounts of human nature that might guide social policy in any global fashion.

These examples concern research on behavior where the multiplicative strategy may be socially and pragmatically appropriate. Other areas of research may require different ways of handling the possibility of pluralism. For example, estimates of the strength or degradability of materials used in nuclear waste storage facilities cannot be left in a state of uncertainty. Conversely, however, the consequences of acting prematurely without considering the variety of frameworks within which estimates might be generated could be catastrophic. Scientific advisory panels and granting agencies cannot proceed as though there will be just one correct account of a phenomenon under investigation, thus taking a string of empirical successes as proof of that correctness. They must instead incorporate the possibility of plurality into their decision-making procedures.[7] How this should be done is a topic too broad to address here. Richard Rudner's analysis (1953) would not be a bad place to begin one's inquiry.

Conclusion

The ideal of value freedom was advanced because it was thought that value-free science could best ensure impartial (unbiased, socially neutral) science and universally valid science, that is, results that would hold for anyone, anywhere. This has led individual investigators to suppose that they must keep their own values out of the laboratory and that doing so would be sufficient to guarantee value-free, impartial science. I contend that the conception of inquiry this thought is based on is untenable and furthermore that the values held by the entire community will not be checked by vigilance for the idiosyncratic. The alternative, social, account of knowledge indicates that the objectives of the value-free ideal are better achieved if the constructive role of values is appreciated and the community structured to permit their critical examination. Structuring the community to include multiple perspectives and values will do more to advance the aims in relation to which value-free science was an ideal — impartiality and universality — than appeals to narrow methodology ever could.

NOTES

1. Of course, one might take an alternative view and argue that what the sciences proclaim about human differences should have no bearing on social policy, that such policy ought to be determined by our political goals and values and not be composed of transient empirical theories. I agree that there is a good argument to be made for this conclusion, but I do not think this precludes an investigation into the grounds for the claims of scientific value freedom.

2. I use *cognitively rational* and *cognitive rationality* to distinguish the kind of rationality in question here from pragmatic rationality, which is not understood as excluding the social in the same way.

3. One factor contributing to the confusion is ambiguity of the word *social*. It is used to refer to human relations, activities, and interactions; to the content of both normative and descriptive propositions; and to the shared character of some content.

4. My formulation here deliberately equivocates between an ontological and an epistemological articulation.

5. This is to say, not that scientists face a gap over which they leap with careless abandon, but that the ways in which the gap between hypotheses and data is closed involves reliance on assumptions that are contestable.

6. Using this social account of justification one might then say that some content A (a theory, model, hypothesis, observation report) is epistemically acceptable in community C at time t if A is supported by data d evident to C at t in light of reasoning and background assumptions that have survived critical scrutiny from as many perspectives as are available to C at t, and the discursive structures of C satisfy the conditions for effective criticism. In Longino (2002a, 135–40), I use this notion of epistemic acceptability to provide accounts of epistemological concepts.

7. For debate on this matter, see the exchange between Philip Kitcher and myself (Kitcher 2002a,b; Longino 2002b,c).

REFERENCES

Barnes, Barry, and David Bloor. 1982. Relativism, rationalism, and the sociology of scientific knowledge. In *Rationality and relativism*, ed. Martin Hollis and Steven Lukes, 21–47. Oxford: Blackwell.

Bloor, David. 1991. *Knowledge and social imagery*. 2nd ed. Chicago, IL: University of Chicago Press.

Collins, Harry, and Trevor Pinch. 1993. *The Golem*. Cambridge: Cambridge University Press.

Duhem, Pierre. 1954. *The aim and structure of physical theory*. Translated by Philip Weiner. Princeton, NJ: Princeton University Press.

Dupre, John. 1993. *The disorder of things*. Cambridge, MA: Harvard University Press.

Ereshevsky, Marc. 1998. Species pluralism and anti-realism. *Philosophy of Science* 65 (1):103–20.

Goldman, Alvin. 1995. Psychological, social, and epistemic factors in the theory of science. In *Proceedings of the 1994 biennial meeting of the Philosophy of Science Association*, ed. Richard Burian, Mickey Forbes, and David Hull, 277–86. East Lansing, MI: Philosophy of Science Association.

Gould, Steven Jay. 1986. *The mismeasure of man*. New York: W.W. Norton.

Hubbard, Ruth. 1979. Have only men evolved? In *Women look at biology looking at women*, ed Ruth Hubbard, Mary Sue Henifin, and Barbara Fried, 9–35. Cambridge, MA: Schenkman.

Kitcher, Philip. 1993. *The advancement of science*. New York: Oxford University Press.

———. 2002a. The third way: Reflections on Helen Longino's *The fate of knowledge*. *Philosophy of Science* 69 (4):549–59.

———. 2002b. Reply to Longino. *Philosophy of Science* 69 (4):569–72.

Knorr-Cetina, Karin. 1983. The ethnographic study of scientific work. In *Science observed*, ed. Karin Knorr-Cetina and Michael Mulkay, 115–40. London: Sage.

Latour, Bruno. 1987. *Science in action*. Cambridge, MA: Harvard University Press.

———. 1993. *We have never been modern*. Cambridge, MA: Harvard University Press.

Laudan, Larry. 1984. The pseudoscience of science? In *Scientific rationality: The sociological turn*, ed. James Brown, 41–73. Dordrecht: Reidel.

Longino, Helen E. 2001. What do we measure when we measure aggression? *Studies in History and Philosophy of Science*. 32 (4):685–704.

———. 2002a. *The fate of knowledge*. Princeton, NJ: Princeton University Press.

———. 2002b. Science and the common good: Thoughts on Philip Kitcher's *Science, truth, and democracy*. *Philosophy of Science* 69 (4):560–68.

———. 2002c. Reply to Kitcher. *Philosophy of science* 69 (4):573–77.

———. 2003. Behavior as affliction: Framing assumptions in behavior genetics. In *Mutating concepts, evolving disciplines: Genetics, medicine, and society*, ed. Rachel Ankeny and Lisa Parker, 165–87. Boston, MA: Kluwer.

———. Forthcoming. Theoretical pluralism and the scientific study of human behavior. In *Scientific pluralism,* ed. C. Kenneth Waters, Helen E. Longino, and Steven Kellert, Minneapolis: University of Minnesota Press.

Mitchell, Sandra. 2002. Integrative pluralism. *Biology and Philosophy* 17:55–70.

Pickering, Andrew. 1984. *Constructing quarks.* Chicago, IL: University of Chicago Press.

Rosenberg, Alexander. 1994. *Instrumental biology or the disunity of science.* Chicago, IL: University of Chicago Press.

Rudner, Richard. 1953. The scientist qua scientist makes value judgments. *Philosophy of Science* 20:1–6.

Shapin, Steve. 1994. *The social history of truth.* Chicago, IL: University of Chicago Press.

Waters, C. Kenneth. 1991. Tempered realism about the force of selection. *Philosophy of Science* 58 (4):533–73.

Witelson, Sandra. 1986. An exchange on gender. *New York Review of Books* 32 (16):53–54.

8

"Social" Objectivity and the Objectivity of Values

Tara Smith
Department of Philosophy, University of Texas at Austin

In recent decades, the longstanding image of science as value free and objective has come under attack. Part of this attack is justified. For scientific inquiry is saturated with value judgments. In determining what to study, what to look for, which methods to employ, which hypotheses to investigate and which to dismiss, how many studies to conduct to confirm initial findings, and so on, scientists are invariably discriminating between better and worse ideas. However internal or external the values employed — whether they are standards developed within biology for testing hypotheses, for instance, or values beyond science, such as the desire to combat a rising incidence of breast cancer — reliance on value judgments seems inescapable. Science itself is the pursuit of knowledge — a quest for value — and as such must reject those questions and methods that are not fruitful.

Contrary to an increasingly popular inference, however, this immersion in values need not compromise science's objectivity. The belief that it must rest on the presumption that values are not objective. This assumption is what I wish to challenge. I can only discuss aspects of that challenge here. Demonstrating that values are objective requires a clear understanding of values and of objectivity, both of which are large and complex subjects. My aim here is not to offer a definitive account of either but, more modestly, to indicate ways in which objectivity is often misrepresented and, more briefly, the manner in which values stem from the same facts that underwrite the objectivity of knowledge in other realms.

In considering the nature of objectivity, I will focus on critiquing the doctrine that objectivity is "social." Since this influential idea distorts expectations of what an objective account of knowledge or of value should look like, such a critique is needed to clear the ground for a more accurate grasp of objectivity. In considering value, my focus will be more positive. Alongside a number of moral philosophers who have recently tendered naturalistic accounts of value, I will sketch the natural fact-based explanation of value proposed by Ayn Rand, according to which values must be objective in order to fulfill their function of enabling human beings to live.[1]

The Social Conception of Objectivity

In light of the work of Thomas Kuhn (1962), Paul Feyerabend (1978), W. V. Quine (1953, 1960, 1969), and others, the objectivity of science has, for some time now, been on the defensive. Rather than despair of the possibility of objectivity, some have urged a social understanding of what constitutes an objective method. I will take as my exemplar of this view Helen Longino.[2] Longino observes that science is an intricate collaborative effort. Especially in our era of "big science," scientific research is a social enterprise in which numerous individuals contribute pieces of larger puzzles, with many people offering hypotheses, experiments, variations, and criticisms. Individuals build on one another's work, as knowledge is generated through "critical emendation and modification" (Longino, 1990, 67, 68). To guard against researchers' prejudices, funding and publication are determined by peer review—the eyes of others (68). It is exactly this social character of inquiry, Longino maintains, that protects science from "unbridled relativism" (216, 221).

According to Longino, "Only if the products of inquiry are understood to be formed by the kind of critical discussion that is possible among a plurality of individuals about a commonly accessible phenomenon, can we see how they count as knowledge rather than opinion. (Longino, 1990, 74; see also 73). A few pages later, "The objectivity of individuals . . . consists in their participation in the collective give-and-take of critical discussion and not in some special relation (of detachment, hardheadedness) they may bear to their observations (79). Yet again, "The greater the number of different points of view included in a community, the more likely its scientific practice will be objective, that is, that it will result in descriptions and explanations of natural processes that are more reliable in the sense of less characterized by idiosyncratic subjective preferences of community members (80). Longino maintains that, other things being equal, a more inclusive theory is better than a less inclusive theory, "not as measured against some independently accessible reality but better as measured against the cognitive needs of a genuinely democratic community" (1990, 214).

Longino is certainly not alone. Sandra Harding (1991, ix) contends that science and knowledge are created by "contemporary social relations," though she, too, retains the language of objectivity. In Harding's view, genuine objectivity is achieved by the democratic incorporation of multiple standpoints (cited in Haack 1998, 116).[3]

Steve Fuller contends that "the communicative process itself is the main source of cognitive change" and writes that all epistemology worthy of the name is motivated by essentially sociological considerations (Fuller 1988, xiii, 7). Stephen Jay Gould declares that science is not an objective enterprise, but a "socially embedded activity ... The most creative theories are often imaginative visions imposed upon facts; the source of imagination is also strongly cultural" (Gould 1981, 21–22). Marjorie Grene emphasizes scientists' reliance on others for education, support, ideas, instruments (Grene 1985, especially 13–16). Kenneth Gergen goes so far as to say that "the validity of theoretical propositions in the sciences is in no way affected by factual evidence" (Gergen 1988, 37).[4]

What's wrong with this picture? Essentially, what social objectivity delivers is merely a bridled relativism. At core, it is no more objective than the unbridled relativism that Longino wishes to avoid. To see this, consider: why is the social perspective presumed to be more objective than an individual's perspective? Why is the fact that a number of people agree on an idea any assurance that *what* they agree on is objectively valid? Now, of course, it is not quite that simple; Longino envisions "critical" discussion and "critically achieved consensus" (1990, 74, 79). The problem, however, is that she gives us little account of the standards that are to govern these discussions. What will render them sufficiently "critical" to confer objectivity on their conclusions? Should the group ask the right questions? Are some questions better than others? Are *any* criticisms or any hypotheses that group members raise valid, simply because *they* raised them? Surely the answers to these questions are crucial if the results of such "democratic dialogue" are to be christened "objective." Longino no doubt intends "critical" and "democratic" consensus to exclude certain considerations and procedures. And in her more recent work, she has said more about the conditions of community discussion (2002, 128–35). Yet her requirements of "public standards that regulate discursive interaction" and "tempered equality of intellectual authority," among other conditions, remain too vague to be very helpful (Longino, chapter 7, this volume).

On contemplating these questions, we realize a deeper difficulty. Any adequate account of objectivity must be based on a standard external to a group's agreement. Failure to recognize this is a fatal flaw. The defense of the social conception of objectivity will be plausible only to the extent that it relies on a nonsocial standard of objectivity. Notice that advocates of social objectivity do not simply enumerate their ranks as the reason why others should accept their view. Their argument is not that Tom, Dick, and Harry (highly respectable sorts) are social objectivists; therefore, you should be one, too. Rather, they recruit independent evidence on behalf of their position — maintaining, for instance, that science benefits from collaboration and mutual criticism. It is that benefit that lends credibility to their position, however, and not credibility per se. But what is responsible for that benefit? Indeed, how could we distinguish certain effects as benefits without tacitly assuming that certain states of affairs are objectively superior to others? If all that a "benefit" referred to was the fact that social objectivists accepted a particular view, their argument would be hopelessly circular: this method is better because it's beneficial — that is, because we all agree it is better. To avoid that circularity, however, social objectivists must contradict themselves, simultaneously holding that the test of a claim's objectivity is the agreement of certain people and that the test of a claim's objectivity is not that, but certain facts independent of their agreement.

The point is this: the appeal of the social conception of objectivity (to anyone other than a previously committed social objectivist) depends on reaching beyond purely social standards of objectivity and embracing an external standard, thus defying its own claim. Obviously, such an external standard can only be known through individuals' observations, but this does not entail that those observations themselves become the substance of objectivity, or that what renders a claim objective is people's agreement about their observations. Yet that is what the social conception of objectivity maintains. (Recall Longino's claim that objectivity *consists in* people's participation in critical discussion.)

As a matter of the sociology of science, a theory's failure to mesh with the dominant theories in a field may well count heavily against its acceptance. As a matter of rationality, however, this is hardly a refutation of that theory. For those dominant theories might themselves be mistaken. (If the advocates of social objectivity insist that a conclusion is not knowledge until it is socially endorsed as knowledge, they again beg the question: assuming that objectivity is social, rather than showing that it is.) History has provided abundant cases of the expert consensus being overthrown by the ideas of a lonely dissenter.[5] It is true, as a matter of history, that dissenters could only succeed in overthrowing a previously entrenched theory because their novel theories won approval from relevant individuals. When such acceptance marked genuine progress in scientific understanding, however, the grounds for that acceptance rested in observed phenomena and the plausibility of the dissenter's explanation of the phenomena. If the advocate of an unorthodox theory were to appeal merely to scientists' opinion, he could never budge the status quo. Rather, the dissident must point to reality, through evidence and rational inferences, to sway other scientists to surrender the entrenched view. A consensus around a newer theory may well eventually follow its presentation and defense, but such a consensus is not what renders it a better or more objectively valid theory.

Contrary to the social objectivists, the antidote to subjectivism in science is not more people, but more rational people. The more of those, the better — better *because* they give us greater chances of uncovering facts.

And this points to a still deeper problem with the social conception. It misses the heart of objectivity: fidelity to reality. Objectivity demands exclusive adherence to those aspects of reality relevant to any inquiry and a scrupulously logical identification of what the observed aspects signify. Ultimately, it is the world — the way things really are, as we often say, independent of our beliefs — that human beings must navigate in order to succeed in life, rather than simply some community's beliefs about the world.[6]

It is true that a scientist's theories are often corrected by others, build on others' work, and develop in collaboration with others.[7] That does not mean that what we need to know is simply the current scientific consensus about some phenomenon. What we need to know is the way things are: What *is* the nature of this tumor? The effect of this treatment? The power of that bomb? My point is not that objectivity requires omniscience or infallibility. As the product of fallible beings, the best of our scientific knowledge today may be significantly altered tomorrow. Nonetheless, the appropriate aim of scientific inquiry is the discovery of reality.

What is alluring about the social conception is that it begins by making true observations about the frequently salutary effects of cooperative inquiry. It thus acquires the air of being realistic. What is fallacious, however, are the conclusions drawn from these observations. Peer review, for instance, undoubtedly can be an effective check against bias. This is so, though, because it can uncover actual errors in a researcher's findings or reasoning. Peer reviewers' scrutiny is useful, in other words, to the extent that it is based on reality. It is not the sociality of the practice that holds value, but the possibility of a given reviewer catching factual oversights, discrepancies between observations and conclusions, significant omissions, logical gaps, or (based on evidence and rational chains of inference) seeing promise in a path not previously pursued, or evidence for a new conclusion. The reports of an assortment of sloppy, subjective peer reviewers do not add up to an objective evaluation.

Knowledge certainly does often benefit from the contributions of many people. It does not follow from this, however, that others' beliefs should become the standard of objective knowledge. Yet this is the implication of the social conception, which replaces the relevant reality as the arbiter of knowledge with a group's "critical consensus." That is not a sufficiently fundamental touchstone for objectivity, however. It is inadequate because believing, like wishing, does not make things so.[8]

Advocates of the social conception might object that I am misrepresenting their view. They do not mean to replace reality with consensus in steering the quest for knowledge; they merely mean to acknowledge the important role of social input. The insistence on *critical* consensus is meant to guard against unqualified deference to a group's opinions. In chapter 7 Longino contends that the lesson of underdetermination is that we need an "expansion" of our understanding of justification, rather than its rejection.

The problem, however, is that the proposed expansion shifts the standard for assessing objectivity. That is, if we expand our understanding of justification, we still need to know what has final say in deeming a theory objective. What serves as the ultimate test, within the reconfigured model of justification? In their eagerness to stress the social dimension of scientific inquiry, advocates of the social conception diminish the crucial role of reality—of facts about the world that should constrain people's theories. *Of course* it is human beings who interpret observations and formulate theories to explain phenomena. In order for such theorizing to arrive at objective knowledge, however, such actions must be faithful to the way things really are. That must remain the fundamental touchstone of objectivity. Longino's contentions that objectivity *consists in* participation in discussion with others (1990, 79), that knowledge requires discussion (74), and that theories are to be evaluated not by the yardstick of the reality they purport to describe but by the cognitive needs of people (214) fail to recognize this.

As advocates of the social conception go, Longino is a moderate; her views are much more tempered than those of many others. Longino wishes to reconcile "the objectivity of science with its social and cultural construction" (Longino 1990, ix) and seeks to avoid radical relativism. It is precisely for this reason that I take her as the representative. For even on this moderate version of social objectivism, science *is* social knowledge (as the title of her 1990 book *Science as Social Knowledge* indicates and as her 2002 book reaffirms, ix), the objectivity of science is *secured* by the social character of inquiry (Longino 1990, 62, 216), and the objectivity of science is a *consequence* of scientific inquiry's being a social rather than an individual enterprise (67). In these characterizations, I submit, the social conception of objectivity strays from objectivity's foundation in reality, embracing conclusions that are not warranted by its observations about human beings' role in science. Essentially, social objectivists seek to have it both ways: they desire all the respectability of reason alongside all the wiggle room of social agreement. But *both* cannot be the ultimate arbiter of objectivity.

In Longino's view, it is divergent values that drive the expansion of knowledge (chapter 7). If you do not like the results of a study finding correlations between race and IQ, for instance, she recommends that you pursue alternative explanations of the findings (chapter 7). Does this truly drive an expansion of *knowledge,* however? Not necessarily. It can, when the alternative explanation turns out to be valid. Yet often the explanation will not be, and this deference to values will merely invite a proliferation of conflicting theories tailored to appeal to specific constituencies. "All those who believe in god, here's your theory of the origins of the universe. All those who don't believe in god, here's your theory." This betrays the idea that science is to describe the *world*. Rather, it depicts science as offering theories of the world as a particular group would like it to be or as fit comfortably with some people's preexisting conclusions.

To be clear: I am not opposing the idea that values play an important, often constructive role in spurring the growth of scientific knowledge. I am opposing the idea that social consensus should determine what should be respected as objective. The immediate point is that, given any expansion of our concept of the justification of knowledge claims, we will still need to identify the decisive barometer of objectivity. If it is inquirers' preexisting values, or the agreement of others, then reality *has* been replaced in serving that role.[9]

Metaphysical and Epistemological Objectivity

An advocate of the social conception might try another tack in responding to my argument thus far. He might protest that he appreciates reality's independence. Nonetheless, he insists, individuals by themselves are ill-equipped to discover its nature; only a suitable exchange of ideas can compensate for individuals' shortcomings—and can warrant respect as objective.

This contention cannot salvage the social conception, however. The distinction being tacitly invoked, between metaphysical and epistemological senses of "objectivity," is certainly valid. Metaphysical objectivity refers to an existent's independence from observers' thoughts or wishes. To say that X is objectively so in the metaphysical sense means that X is what it is (exists, possessing the characteristics that it does) regardless of anyone's opinions or attitudes toward X. Epistemological objectivity, by contrast, refers to the method of thinking by which a person could become aware of such existents. Metaphysical objectivity concerns *what* we know (the object of knowledge), whereas epistemological objectivity concerns the means by which we come to know it.

Now, while social objectivists might wish to retain metaphysical objectivity, claiming that they recognize our inability to wish facts into submission, logically, their position on epistemological objectivity precludes it. For the argument for social (epistemological) objectivity can succeed only if metaphysical objectivity is false. To see this, bear in mind that the purpose of epistemological objectivity is to gain knowledge — that is, to learn the nature of reality. If metaphysical objectivity is accepted, then others' opinions (which are pivotal to the social conception) make no difference to the reality that a person is trying to discern. Since thoughts by themselves do not create or alter existents,[10] the quest for consensus is a diversion from the route that knowledge requires. Even the weaker demand for dialogue with others is, at root, beside the point. A critical exchange with others can be helpful, many times, but it is not essential for objectivity and it is not what objectivity consists in.

Still, one might suspect that I am failing to fully appreciate the difference between the two kinds of objectivity. Why could Longino and her allies not simply say, "Of course wishing does not make things so; we certainly would not dispute that. Things are the way they are independently of people's beliefs or desires. All we are saying is that we are unlikely to ever know the way things are by solo efforts. Individuals' perceptions and thoughts are infected by subjective factors that prevent them from offering the kind of accurate awareness that we are after. Dialogue with others provides the corrective necessary to allow for objectivity."

Thus stated, the self-contradiction inherent in this position should be apparent. If the social conception is correct that "we are all infected," then how can social objectivists know the nature of reality? How can they affirm existents' mind-independence as if it were objectively so and not merely the product of a suitable consensus?[11] Note that the latter would not be enough for social objectivists to truly endorse metaphysical objectivity. If all their professed allegiance meant was that metaphysical objectivity holds because it has passed the test of "critical consensus," they would not agree with those who affirm metaphysical objectivity apart from anyone else's beliefs about it. Indeed, they would once again be assuming the subjectivist social outlook, by giving the social interpretation as all that the assertion of metaphysical objectivity *could* mean.

The basic problem is that recognition of metaphysical objectivity requires epistemological objectivity. Social objectivity — insofar as it is not accountable to the facts it purports to describe — could not inform us about the nature of existents. If our best, "objective" thinking reflects nothing more than a consensus that is not required to meet any further reality-grounded standards, on what basis could one claim that things actually do possess a certain character independently of our thoughts? By what route is such a conclusion reached let alone warranted?

If, alternatively, a person claims that consensus alone is not decisive, but that accurate identification of facts is (accurate, that is, as far as presently available methods can determine), this person is abandoning the thesis of social objectivity. And if a person wishes to maintain merely that critical consensus is necessary for the objectivity of a claim but not sufficient, we are left wondering why it should be necessary. To recognize that rational discussion with others can often be extremely valuable to acquiring objective knowledge does not entail that discussion with others or the agreement of others is across the board, in principle, necessary for objective knowledge.

If metaphysical objectivity obtains and existents' nature is independent of observers' thoughts and wishes, then only certain methods can succeed in informing a person about the nature of reality. When we speak of the mind-independent nature of reality, this encompasses the nature of human thought; the brain has a definite nature and a person has limited, specific means of acquiring knowledge. Individuals' sensory and intellectual equipment determine the kinds of observations that they can make and the thought processes that they can engage in. Their own natures and the natures of what they observe together entail that only certain ways of using their observational and intellectual capacities can yield knowledge. Securing the agreement of other people is not one of them. A genuine recognition of metaphysical objectivity requires an epistemological objectivity, an appropriate method of using one's mind, that is not wedded to others' opinions. Unless a person uses the methods for acquiring knowledge that the metaphysically objective nature of reality demands, this person cannot know anything about reality — including its metaphysical independence.

Underneath the errors of the social conception, I think, lie confused expectations of how knowledge should be available and what objectivity should deliver. To appreciate these, it will be useful first to say a bit more about what objectivity is.

Epistemological objectivity concerns a person's manner of using his or her mind. (Henceforth, I will be speaking of epistemological objectivity unless I clearly indicate otherwise.) It consists in thinking and drawing conclusions based on strict logical adherence to the relevant facts. By this I mean that a person's inferences from his or her observations are disciplined by a rational understanding of the relationships among facts. The objective person does not leap to sweeping or unwarranted conclusions on the basis of paltry evidence, for instance. A man, say, would not conclude that his wife was having an affair simply because she is late coming home from work one night; he would not conclude that a particular person is untrustworthy simply because another person of the same race was untrustworthy.

The purpose of a specific inquiry, along with knowledge from past experience, will determine what information is relevant, as well as the relative weight to be assigned various factors. The objective teacher, for instance, grades on the basis of an essay's quality rather than on a student's need for a particular grade to maintain a football eligibility. The teacher judges that quality by reference to such factors as the purpose of the assignment, the educational level of the student, the relevant material taught in class, the paper's clarity, organization, the pertinence and thoroughness of the points offered to support its thesis. We can formulate these factors more precisely in some areas than in others (for example, in math versus in literature), but even where there may be a range of equally acceptable answers, it is a limited range, and answers falling outside of it, failing to make essential points, are inferior. While I cannot offer an extended account of the criterial requirements for grading or any particular activity here, most centrally, logic demands that a person seek and forthrightly identify the relevant facts, integrate them with other knowledge, strive to avoid contradictions in reasoning, and consider the context to properly interpret the significance of individual points.[12]

To be objective, as the word's Latin roots imply ("ob," in front of; "jacio," to throw), is to focus on and be faithful to that which is thrown in front of you. Objectivity is a deliberate commitment to keeping one's beliefs grounded in reality by thinking logically. Objectivity is a method for human beings—who are fallible—to employ to discipline our thinking to help us gain an accurate understanding of the world.[13]

A few aspects of objectivity, because they are often misunderstood, bear special attention. First, being objective does not extinguish a person's fallibility. That the objective method places us in the best position to gain knowledge does not mean that all persons employing it will arrive at the same conclusion or at the correct conclusion. Individuals' differing knowledge and differing intellectual capacities allow for such variations. (Also, a single correct conclusion does not always exist; two or more solutions to a problem are sometimes equally rational.) Nor does the use of the objective method mean that answers reached by that method will invariably endure for eternity. The contrary expectation is commonplace. A typical example is Antonio Damasio's breezy portrait in the introduction to his book *Descartes' Error: Emotion, Reason, and the Human Brain*: "I am skeptical of science's presumption of objectivity and definitiveness. I have a difficult time seeing scientific results . . . as anything but provisional approximations, to be enjoyed for a while and discarded as soon as better accounts become available" (Damasio 1994, xviii). His implication is that objectivity entails immunity to revision.

In fact, further knowledge uncovered in the future may reveal that an objectively valid answer today is not complete, or is not accurate, in the context of additional information. Thus a person could make a mistake while being objective. A jury, for instance, could scrupulously weigh all testimony and evidence heard at trial, reason impeccably about every relevant aspect of a case, yet still reach an incorrect conclusion about a defendant's guilt because of facts that only later came to light and which the jury had no reason to suspect. The error would reveal that the jury was not objective, however, only if one supposed that being objective should somehow transform us into error-proof beings. This is an unrealistic image of what objectivity is and does.[14] (Note that "critical consensus" offers us no immunity from sometimes having to make the same sort of revisions. With social objectivity, however, shifting opinion would prompt revisions rather than the discovery of further facts. Such shifting opinion might be based on discovery of facts — but it might not be, and it does not need to be.)

Second and closely related, objective knowledge is not akin to revelation. The assertion that "X is objectively true" is not a claim to special insight, to be among the chosen to whom some eternally incontrovertible fact has been revealed. Nor is it a claim to enjoy a god's perspective or a "view from nowhere." No such perspectives are available to human beings. It is because we are fallible and potentially subjective that we need to discipline our thinking to make it accurately grasp reality. Objectivity is a method of doing so, but it is a method *for human beings*. As such, an understanding of what it is and offers must be based on our actual capacities. Although a person claiming that X is objectively true *is* claiming that X is so, this person should be understood as saying that X is so insofar as the most reliable methods at present can discern. While being objective will strengthen the reliability of our conclusions, it cannot convert us into infallible or omniscient beings, magically capable of knowing things from all conceivable perspectives, no longer limited by the nature of our perceptual and mental equipment. A rational concept of objectivity will not require us to function as some fictional kind of being.[15]

What animates the social conception of objectivity seems, in large part, the essentially Kantian idea that we can never know the world as it really is. As Harding puts it, "Nature as-the-object-of-human-knowledge never comes to us 'naked'" (1991, 147). Advocates contend that belief in even the least controversial tenets of science relies on a person's perceptual apparatus, background beliefs, interpretations. Because direct, unfiltered access to reality is impossible, the best that objectivity could be is a group's agreement.

Here, too, however, the social conception is operating with skewed expectations. To seek a transparent vision of reality, some sort of unfiltered insight, is to seek magic: knowledge attained by no specific means. In fact, a process of acquiring knowledge, with the requisite equipment and activities on the knower's part, is ineliminable. To know anything, a person must know it *somehow*. Far from compromising the objectivity of knowledge, it is such mechanisms and activities that make objectivity possible. As Ayn Rand has observed, no one would conclude that because a stomach must digest food we can have no objective standards of nutrition. It would be equally ludicrous to conclude that because human beings must utilize certain resources and exercise certain intellectual processes to gain knowledge we can have no objective standards of knowledge (or that our conception of objectivity must be socialized) (Rand 1990, 81–82).[16]

The social conception reasons as if acquiring knowledge should be a matter of simply directing one's gaze outward and waiting for reality to write itself onto one's mind. Since, it correctly observes, such imprinting never takes place, it concludes that *we* invent knowledge by pooling our minds.

Unfortunately, this reasoning simply leaps from an erroneous intrinsicist conception of knowledge to a subjectivist one, with genuine objectivity lost in the lurch. Properly, objectivity pertains to the relationship of consciousness to existence, with both elements vital. One cannot bypass reality in order to gain objective knowledge (as a social standard does); nor can one omit the role of the mind — in particular, the need for a person to use his mind in a disciplined, logical way in order to arrive at genuine knowledge. Objectivity resides neither wholly in our heads, independent of external reality (as subjectivists would have it) nor wholly in external objects, independent of our mental processes (as intrinsicists would have it). While the social conception is right to reject the idea that reality simply stamps knowledge onto our minds, it fails to realize that that is not the only alternative to the subjectivism that it attempts to pass off as objective.[17]

While I cannot offer a full explication of objectivity here, nonetheless, let me consider a few final objections before turning to values.

Objections

One might protest that I am propping up the possibility of objectivity (a nonsocial brand) only by dumbing it down, deflating what objectivity means so as to make it more attainable. Thus, even if objectivity is possible, it will turn out to be less valuable. I do not think that this charge can stick, however.

"Dumbing down" means reducing appropriate standards. If we teach easier material in school so that more students can pass, we are dumbing down the curriculum. What I have been urging is converting from unattainable standards of objectivity to attainable ones. That is not dumbing down; it is, in the vernacular, getting real. The assumption that "objective *for us*" means "not truly objective" relies on a standard that is not based on our actual capacities. The point of such a standard is hard to see.

If objectivity is to serve as a useful ideal, it must be based on our actual capacities. Given that we are fallible and capable of prejudice and partiality, the question is, are there more and less reliable means of trying to obtain knowledge, such that some methods should be encouraged as objective and others discouraged as subjective? In answering yes, while recognizing what objectivity cannot do, I am not diminishing objectivity; I am understanding it as an ideal that we really can practice and, consequently, as one that we really should. Without this belief in its possibility, any injunction to be objective rings hollow (since "ought" implies "can"). No benefits can accrue from telling people they ought to practice an ideal that they are incapable of practicing. Far from my conception's proving less worthwhile, it is only such realistic objectivity that can truly assist people in acquiring knowledge.[18]

A second objection is more serious, more far-reaching: my account is naive. The contention that objectivity is a matter of fidelity to reality or "following the facts" neglects the subjectivizing influences on anyone's understanding of what the facts *are*. Many critics of objectivity have claimed that individuals cannot escape personal biases, however hard we might try. Longino writes that the belief that we can "just read the data" is a recipe for replicating mainstream values and ideology (Longino 1990, 218). Harding contends that "What the [natural] sciences actually observe is always already fully encultured" (92).[19]

In response, I would make two points. First, how could I refute this charge? I cannot. I cannot, *not* because the accusation makes a valid point which has been shown to be true, but because it will not allow anything to count as refutation. What would qualify as counterevidence? A researcher's thoroughness? Adjusting his inferences to compensate for factors potentially prejudicing his findings? The corroboration of studies conducted independently, in different cultures (thus ensuring diversity of inquirers)? These efforts would not satisfy the critic. Anything offered could simply be reinterpreted as itself infected by the same plague of prejudice. The thesis that "all readings of data are biased" is a conveniently self-serving theory that disarms any attempt to challenge it. It "wins" not by demonstrating fact, but by fiat, by arbitrarily insisting that nothing but it can win.[20]

It is true that theories do not construct themselves; data do not literally put words into our mouths or thoughts into our heads. People must interpret data to make claims about the world. And it is in doing so that the possibility of bias arises. The thesis charging the inescapability of bias, however, maintains that this ensures the certainty of bias. That is its mistake.

It is because individuals *can* allow subjective factors undue influence on our thoughts that it makes sense to employ the objective method. By proceeding objectively, by deliberately tying our claims to our observations and to scrupulously rational inferences built on those observations, we minimize chances of error. Objectivity is a tool to help bind what we think is so to what actually is. The fact that we must exert choices in using the tool does not render our conclusions subjective. On the contrary, it is only the use of this method that permits our escape from subjectivity.

The "no escaping bias" objection implies that knowledge, to be objective, should be acquired effortlessly, through a mysteriously means-less process, like the passive receipt of revelation. The objective person is envisioned as a programmed robot that can exert no choices. For once a person has options, the "no escape" thesis holds, objectivity is doomed (since people will allegedly exercise those options on the basis of subjective factors).

Yet this view, once again, presumes a distorted model of what objectivity is and provides. It is not the fact of choices that renders a conclusion subjective but which choices a person makes—which considerations he treats as relevant and what weight he assigns them in his thinking. Objectivity is not precluded by a person's need to direct his thinking. To think that it is is comparable to suggesting, as Rand puts it, that "man is blind, because he has eyes—deaf, because he has ears" (Rand 1961, 32).[21] Choice is essential to thinking; it enables us to think rationally. The point of methodological objectivity is to guide a person's choices so that they lead to knowledge. My second point in reply: objectivity is neither rescued nor strengthened by appeals to society.

Longino's charge is that "reading the data" is inevitably distorted by subjective values. This raises a serious question: how does her account of objectivity escape this problem? Recall that for Longino objectivity is secured by a community's critical discussion (Longino 1990, 79). The greater the variety of perspectives included, she holds, the more objective, because the scrutiny of a large and diverse group is likely to expose the "idiosyncratic subjective preferences" of particular individuals within that community (1990, 80). Subjecting ideas to group debate cleanses them.

Yet, I wonder . . . We could only expose a given individual's view as idiosyncratic or subjective against some standard. What is the standard that Longino is employing? Others' opinions. For all the references to critical discussion, Longino never makes clear what constitutes a suitably critical dialogue or exactly how that is to rein in the conclusions of a group of people. By requiring discussion for the acquisition of knowledge, by expressly rejecting the thesis that objectivity rests in a person's relationship to his or her observations, and by explicitly disavowing a theory's relationship to reality as the standard for evaluating that theory (Longino 1990, 74, 79, 214), Longino suggests that the ultimate standard lies in the views of other people. But why are the views of those criticizing some individuals as idiosyncratic not themselves idiosyncratic? How do we know that *they* are not subjective? Or simply subjective prejudices that happen to be shared by a number of people? How does sharing a bias make it any less of a bias? By Longino's logic, a position's objectivity is fluid, determined wholly by the number of others who happen to agree with it at a particular time. As popular sentiment ebbs and flows, so does objectivity.[22]

It is one thing to say that an idea will be *considered* objective when a number of people accept it; what I object to is saying that it actually *is* objective.[23]

Still, one might ask, is it not useful to know when a convergence of opinion forms around some conclusion? Is that not a strong indicator of that conclusion's merit? Undoubtedly — sometimes. It depends on whose opinions are converging. If people you know to be rational, independent thinkers agree about some conclusion, that is a positive sign about its validity. It is not proof, but a track record of responsible thinking is some reason to take seriously what several such people are concluding (assuming they are intelligent and informed on the issue in question). Notice, though, that that reason remains grounded in more basic facts about how they form their opinions. If you learn simply that a bunch of conformists agree on something — or a bunch of cultists or flagrant irrationalists, such as Holocaust deniers — that gives you no reason to give credence to their conclusion.

To return to the immediate problem: Longino has provided no escape from her contention that each community member's "readings of data" are contaminated by his or her own particularity. She has offered no independent basis for thinking that one member's view is any more reliable than another's. Beth's view is simply her view, and Mark's is his — each subject to all the contaminating influences that may have prejudiced her or him. Given this, why should Mark's criticism of Beth's view as subjective weigh any more than her charge that his is? By invoking society to settle such disputes — "well, Mark's is the critical consensus" — Longino is offering an analogue to the moral position that might makes right. Longino holds that might makes epistemological right. Numbers dictate truth. That is not objectivity. That is Lysenkoism.[24]

I am not attributing sinister motives to Longino; she herself criticizes Lysenko (Longino 1990, 78). And she is not urging that we force anyone to accept conclusions, as he did. I do mean to point out the alarming implications of "social" objectivity, however. Once one allows that consensus determines objectivity, one is licensing the same sort of "science" that Lysenko peddled. The common underlying idea is, "This is true because we believe it," rather than "because evidence confirms it." Without clear, reality-based constraints on what a group may declare real or true, nothing can stop the group from the brazen defiance of observation that made Lysenko notorious.

While Longino complains that Lysenko's suppression of certain biological views "was a matter of politics rather than of logic or critical discussion" (Longino 1990, 78), having defined objectivity as critical consensus, she is no longer entitled to invoke that distinction. Social objectivity *is* the politicization of knowledge (as its advocates are often eager to broadcast).[25] One cannot deny that objectivity is a matter of logic, or that logic must discipline "critical consensus," and then turn around and invoke logic to criticize politicized science. By the social conception's own standards, a Lysenko does not need facts or logic to vindicate his "scientific" claims. He simply needs enough other people (or, as in Lysenko's case, the people holding political power).

In short, consensus is not the disinfectant that Longino supposes. Numbers offer no immunity from error or from bias. A group of people's sincere conviction about some mind-independent fact does not alter the nature of that fact.[26] It is not enough to cure a man's cancer to learn the medical consensus about effective treatment; a cure demands treatment that really is effective. Metaphysical objectivity, in other words, demands epistemological objectivity. And we have no business calling "objective" a method that turns on agreement rather than on facts.

The Objectivity of Values

Having explored the nature of objectivity, we are in a better position to at least begin to examine the objectivity of values. For even if a robust objectivity is possible, questions remain over the values that scientists rely on in making decisions. It is here, the frequent assumption is, that objectivity is lost, because value judgments are by nature subjective.

The roots of this supposition lie largely in two ideas: the belief that we cannot derive an "ought" from an "is," and the belief that we cannot *reason* about ends. Considering these will lead us to unearth the deeper source of skepticism toward values' objectivity.

Philosophers are taught that logic forbids deriving a conclusion concerning what ought to be from premises concerning what is. Since facts cannot by themselves justify ascriptions of value, objectivity for values is elusive.

Now while our inability to deduce values from facts is unassailable, reason does prescribe, on the basis of induction, better and worse means of accomplishing specific goals. If you want to go to medical school, you ought to study hard in biology; if you want to lose weight, you ought to eat less and exercise regularly. We routinely reason from "is" to "ought" in this way, as I am hardly the first to notice. Certain means of accomplishing ends can be shown to be superior to others.

Here, though, we meet the second apparent obstacle to values' objectivity: How can ends be evaluated? Granted, given particular aims, certain "goods" or "oughts" can be rationally established, but what is the basis for one's ends? Aren't they just a matter of personal choice, of subjective preference?

This, I think, is where opponents of objectivity do not probe deeply enough. We can reason about ends, and we do — largely by reference to other ends. "If you spend all that money on a vacation, you will not be able to finance medical school next year." "If you do not lose weight, you risk developing heart disease."[27]

Obviously, such reasoning can only extend so far. At some point, the chain tracing certain ends' effects on a person's other ends must stop. There, in the choice of ultimate ends, do we not reach the point of no return from subjectivism?

I think not. For not all ends that a person might pursue are equally conducive to his well-being — or to his living. That is a factual, objective basis for judging some ends as better than others.[28] And this leads to the deeper source of skepticism toward values' objectivity.

The real reason that people presume that values are subjective is that people do not understand the function of values. In the sciences and in everyday practical matters, subjectivism is harder to abide because its consequences are so palpable. Faulty brakes, collapsed bridges, or dying babies convincingly refute subjective theories in auto mechanics, engineering, or pediatrics. When it comes to values, however, people are fuzzier about what values are for and are thus uncertain of how to test competing claims about values. If we believe that nothing (or nothing of consequence) hinges on which values a person pursues, we can afford to treat them as subjective. If, on the other hand, we recognize that values have a job to do, we must carefully identify which values will actually perform that job.

I think that values do serve a vital function. And understanding this is crucial to understanding why values not only can be objective but must be. Although I can hardly do full justice to these issues here, let me briefly sketch what I have in mind.[29] A value is that which one acts to gain and/or keep, be it material or spiritual (such as a house, a car, a job, a marriage, freedom, or self-respect) (Rand 1964, 16).[30] Even processes or policies that a person values are things that take some action to promote or maintain; for example, if I value fairness, I will treat others fairly and encourage others to do the same. When we inquire into the objectivity of values, we are concerned with whether a person *should* seek certain ends. It is only life that makes the phenomenon of value possible, however. Nothing can be valuable to inanimate objects. There is no "bad" or "good" to a paperweight, a tire, a rock, or a cave. Events will carry effects on living and nonliving things alike, of course, but such effects can be valuable to an object only when that object has something at stake. Insofar as inanimate objects are not striving to attain any particular outcome or condition, they do not, thus things cannot be valuable *to them*.[31]

Living organisms, by contrast, continually have something at stake: their survival. The fundamental alternative that all living organisms face is life or death. This is an inescapable fact, given by nature. While an inanimate object can be annihilated, destruction would make no difference to it, as it (a paperweight, say) is not trying to do or be anything. Nonliving things are merely passive recipients of events, with nothing invested in different outcomes. Living organisms, however, *face* the alternative of existence or nonexistence because we, typically, are invested in the outcome. While human beings are capable of being the most fully invested, psychologically as well as physically, since we are aware of our struggle to extend our lives and make conscious choices about how to do that, even plants and lower animals are invested insofar as they struggle to live. Living organisms' lives depend largely on their own actions. And this fact carries significant implications for values. It means not only that it is life that makes values possible. Life also makes values necessary. That is, the goal of living demands the pursuit of life-sustaining ends. If an organism is to survive, it must achieve the values that its nature requires.

Herein rests the root of values' objectivity. Life is a series of self-generated, self-sustaining actions. What types of actions are self-sustaining will depend on the nature of the particular organism. Survival requires actions that satisfy the organism's needs. Plants and lower animals are physiologically constituted to act automatically in self-sustaining ways (for example, to absorb water or to hunt), though they are sometimes capable of deviant, self-destructive action. In the case of human beings, however, while some of the actions necessary for our survival are internally wired (digestion and respiration, for example), others must be deliberately performed. Since we choose all of our actions that are not physiologically determined, it becomes imperative that we identify the kinds of ends that will propel our lives and the kinds of actions that will achieve those ends.

Values are objective, then, in that they reflect the factual requirements of our survival. Since our actions carry consequences, we must avoid actions detrimental to our lives and take those actions that preserve them. While it is for human beings to recognize these facts about actions' repercussions and act accordingly, it is nature that dictates what is conducive and what is detrimental to human life.[32]

Because the concept of value makes no sense apart from the struggle to survive and because the desire to survive mandates the achievement of the ends necessary to realize that goal, the standard of value is life. Ends and actions are to be assessed on the basis of their contribution to survival. Obviously, the effects of different ends and actions can be more or less direct, immediate, and profound. While distinctions between good and bad are, at base, a matter of self-preservation, beyond the basics such as food and shelter, things' bearing on human life will be more difficult to determine. (Philippa Foot contends, for example, that industry and tenacity are valuable because they enable people to clothe and feed themselves; 2001, 44–45.)[33] The fact that more abstract values and virtues (types of action necessary for achievement of life-supporting values) require further layers of explanation does not alter the ruling principle, however: that which aids a person's life is good; that which endangers or damages it is bad.[34]

It is also important to appreciate that since it is living that imposes these requirements, satisfying them is valuable only if a person seeks to live. Reality does not issue moral commands; it does not order a person to live (Peikoff 1991, 244). Mind-independent facts demand a certain course of action *if* a person chooses to live. As Rand describes it: "Reality confronts a man with a great many 'musts,' but all of them are conditional; the formula of realistic necessity is: 'You must, if—' and the 'if' stands for man's choice: '—if you want to achieve a certain goal' " (Rand 1982, 99).

Objective value, then, should not be mistaken for intrinsic value.

Value is not an inherent feature of external reality, on my view, nestled within certain things, simply awaiting our discovery. Rather, an objective value is an aspect of reality that stands in a positive relationship to human beings (Peikoff 1991, 241–42). Something is objectively good for a person when it offers a net gain for that person's life. Nothing can be good in itself. A value is always good *to* someone or something, *for* some end (Rand 1964, 16). To say that values are objective is thus to say that the things that people call good and bad stand in factual relationships to individuals' lives, and that there is a (metaphysically objective) fact of the matter as to whether those things are truly life advancing or life diminishing.[35]

Finally, let me warn against another possible confusion. Contrary to its common connotations, "objective" should not be equated with "universal."[36] As we have just seen, values arise only for those who seek to live. While this refers to nearly all people, it does not include everyone. Further, the objectivity of values allows for optional values that can vary considerably among different individuals.

That X is objectively valuable does not mean that it is valuable to everyone. While certain things may be valuable to all people by virtue of their nature as human beings who wish to live (food and shelter, for example), the value status of other things will depend on the relationship of those things to a particular individual in light of that individual's abilities, knowledge, tastes, and the like. Writing philosophy is objectively valuable to me, for instance; it is not to my sister. Accomplishments on the tennis court that would be valuable for a beginner will be far less so for the professional.

Such optional values are not rivals to more basic values; they are the concrete means through which individuals pursue more fundamental values. As a human being, making a living is valuable to Joe. As an individual with particular skills and interests, working as an engineer might be valuable to Joe, but it will hardly be valuable to everyone. Nonetheless, it is objectively valuable to Joe, provided that it serves the role it needs to in advancing his life (at least equally well as alternative work would).

Remember that value is only intelligible in relation *to* some person. Indeed, this is part of what distinguishes objective value from intrinsic value. Value is not a freestanding "good," independent of the role it plays in individuals' lives. Whether something is a value will depend on many characteristics — the full context — of the specific person in question. Thus, variation in people's values per se poses no threat to objectivity. That insulin injections are only beneficial for some people does not mean that they are not objectively valuable (for people who need them). As long as a person's optional values fulfill the requirements of survival, they can be as idiosyncratic as the person holding them yet remain objectively valuable.[37]

All of this has been a highly compressed account of the objectivity of values. In essence, the function of values is to guide human beings to the kinds of ends and actions that we must take in order to sustain our lives. For life hinges on certain needs being satisfied; living requires the achievement of life-sustaining ends. What those ends are is a matter of fact, not of any person's or group's choosing. Individuals' preferences can play some role, as we have just seen with optional values, but that discretion is subject to independent facts about the effects of actions on a person's life. We cannot invent values by personal fiat, because our will does not create the effect things will have on our lives. At core, values reflect facts about survival needs. Correlatively, what underwrites the objectivity of values is that we *need* to be objective in selecting our actions. It is precisely because wishing will not make things so that only the achievement of certain ends will actually advance our lives.

Conclusion

Settling the question of whether values' role in science compromises objectivity clearly depends on understanding the exact nature of both values and objectivity. By pointing to the natural, biological function of values, I hope to have shown that assumptions of values' nonobjectivity are hasty. Indeed, our natural need for values points to their objectivity. Only certain things will fill the life-furthering role that necessitates our identification and pursuit of values.

As for objectivity itself and the prevalent social conception of objectivity, while it is true that science often advances through the contributions of many different people, it does not follow that *what* science should be after — or that what objective knowledge consists of — is simply the consensus of many people. Facts do not bend to a group's beliefs — however well-educated, experienced, diverse, or critical the group. History has shown that science needs its dissidents. Scientific progress has benefited repeatedly from the individual who challenged conventional wisdom. The social conception of objectivity, however, represents the politicization of science. By its standard, what makes a theory true is not, at root, sound research, but effective lobbying—a starkly different enterprise.

The difference between Darwin's theories and those of Holocaust deniers is not simply in numbers of advocates but in kind. Knowledge depends on recognizing that difference. While many of us might wish to protect science from pseudo-science, without the recognition that objectivity reflects something more than a head count we will lack the foundation necessary for doing so.

Without mind-independent facts serving as an essential check on "critical consensus," social objectivity amounts to homogenized subjectivism. But homogenized subjectivism is still subjectivism. And my larger concern is that if we defend objectivity with defective conceptions of what objectivity is, we lose as much as we would if we surrendered to the more radical, "unbridled" relativists. In fact, we probably do worse, because we diminish our understanding of what the alternatives are, allowing things like social agreement to supplant the ideal of fidelity to the relevant facts.

NOTES

I have benefited from comments from audiences at the University of Southern California and the Pittsburgh-Konstanz colloquium at the University of Pittsburgh in October 2002, as well as from discussions with many individuals. I particularly thank the editors, especially Peter Machamer, Mark Bedau, Harry Binswanger, Heather Douglas, Larry Laudan, Jim Lennox, Helen Longino, Greg Salmieri, Wolfgang Spohn, Darryl Wright, and Gideon Yaffe.

1. Some of the other recent authors advocating a naturalistic account of value are Philippa Foot (2001), Rosalind Hursthouse (1999), Berys Gaut (1997), and James D. Wallace (1978).

2. For direct statements of this view, see Longino (1990, 37, 62, 216).

3. Harding speaks of "strong objectivity," claiming that her "feminist standpoint epistemology requires strengthened standards of objectivity," and she expressly wishes to keep the term "objective" so as to develop its "progressive" tendencies (1991, 142, 160–61).

4. Even David Hull, who means to resist relativism and who believes that the natural world should constrain our beliefs, nonetheless describes objectivity as "not primarily a characteristic of individual scientists but of scientific communities" (1988, 3–4).

5. Consider, for example, Galileo, Pasteur, Darwin, Vesalius, Semmelweis, Copernicus, and Harvey. While the dissenters' ideas did not always constitute as radical a break with previous work as popular legend portrays (Darwin was strongly influenced by Charles Lyell's work on the fossil record, for example), these figures were, nonetheless, challenging the prevailing wisdom in their fields. Thanks to Jim Lennox for pointing this out to me.

6. While certain aspects of reality are manmade, such as the Empire State Building and the ingredients of Tylenol, the manmade is itself restricted by the identity of aspects of reality that are not manmade. Only certain materials could be used to construct a durable skyscraper or to create an effective pain reliever.

7. Properly, collaboration is an exchange of ideas for the purpose of enriching the collaborators' ability to gain knowledge. It does not require individuals to surrender their rational judgment to the majority opinion.

8. The error may be a variant of what Susan Haack dubs the "passes for" fallacy. That what has passed for objective truth has sometimes turned out to be no such thing does not entail that the notion of objectivity is humbug, Haack observes — or, I might add, in need of radical reinterpretation (Haack 1998, 93, 117). Haack also observes that the fact that a description involves some conceptual activity on the part of the describer does not mean that it *refers* to the describer's conceptual activity. Yet the social conception of objectivity seems to depend on a comparably faulty inference: since we acquire knowledge through social interaction, it reasons, that must become the object of our knowledge (155).

9. Longino is no doubt correct in claiming that when assessing the degradability of nuclear waste materials, acting before considering different people's different answers could be catastrophic (see chapter 7). That does not show that the proposed answers are all equally objective, however. The means of determining whether they are is reality rather than a political agenda, however widely endorsed.

10. Existents in the external world, that is. Obviously, my thoughts can change certain internal phenomena, such as my plans or evaluations.

11. One might think that I am confusing knowing the nature of reality with knowing that a mind-independent reality exists, which are distinct propositions. Bear in mind, however, that even to affirm the mind-independence of reality is to make a claim about its nature. In other words, these two propositions cannot be entirely separated.

12. For a fuller account of some of the basic requirements of logic, see Peikoff (1991, chap. 4, especially 121–42).

13. This image dominated centuries of scientific progress. William Harvey, one of the pioneers in the gradual objectifying of the science of medicine, described the proper method in very much this spirit: "Nature herself must be our adviser; the path she chalks must be our walk" (quoted in Nuland 1988, 144). Roughly, one might say that the objective person calls it like he sees it and (this part is crucial) strives to see things as they really are. This is only a rough characterization, because it describes perceptual awareness rather than conceptual knowledge, which is the level at which the possibility of objectivity truly arises. It captures the essential idea, however. At the conceptual level of knowledge, the "striving" must be deliberately and carefully guided by logic.

14. Note that the use of instant replay for reviewing calls in professional football and of the "cyclops" electronic eye in tennis is premised not on the presumed corruption of referees but on the same recognition that even well-intended individuals conscientiously trying to get the facts right can make mistakes. Larry Laudan's claim that scientists are not looking for the best theory, but for the best theory they can find is pertinent in this regard (Laudan 1984, 27–28). In one sense, this should go without saying; scientists could find no other than the best theory that *they can find*. Any suggestion that objectivity demands something more than this, such as infallible access to fixed truths, would reflect an unattainable and thus inappropriate image. What is salient to my dispute with the social objectivists is my contention that, properly, scientists should like the best theory they find to *be* the best theory possible, the theory that actually gets the facts right. That is what is lost sight of in the social conception of objectivity.

15. Actually, even a being presumed to be omniscient would have to acquire its knowledge through some means. See a brief but suggestive discussion of this in Rand (1990, 193–94; exchange with Professor F). Thanks to Jim Lennox for pointing out the passage's relevance.

16. Advocates of the social conception do not openly deny objectivity, of course. By redefining it as a subjective phenomenon, however, their stance amounts to the same thing.

17. Rand (1988, 18) proposes this objective/subjective/intrinsic division in regard to values in "Who Is the Final Authority in Ethics?" For a good discussion of the three perspectives in regard to knowledge, see Peikoff (1991, 142–51).

18. The complaint that an ideal is "not realistic" can mean either of two things, not always distinguished: people do not live up to this ideal, or people cannot live up to this ideal. Nothing follows from the first observation alone about what is prescriptively appropriate. My point is that a common conception of objectivity is unrealistic in the second sense, and that any conception of objectivity that is literally beyond our capacities defaults on its task of providing constructive guidance.

19. Further, according to Harding, "Political and social interests are not 'add-ons' to an otherwise transcendental science that is inherently indifferent to human society; scientific beliefs, practices, institutions, histories, and problematics are constituted in and through contemporary political and social projects, and always have been" (1991, 145). The ideal of a disinterested rational scientist simply advances the interests of elites seeking status and power (148).

20. As Peter Machamer has pointed out to me, it also assumes that all enculturation is biased and that conceptualization is the same as enculturation.

21. Rand was speaking of theories of sense perception, but the point is applicable here, as well.

22. Fuller likens epistemic judgment to anticipating trends in the stock market (Fuller 1988, xiii).

23. As I indicated earlier, because of human beings' fallibility and lack of omniscience, even when a conclusion truly is objective, this does not ensure that it will never be overturned. Yet the distinction between sociological observation about which ideas people consider objectively valid and which claims are, at a certain time, actually objectively valid, remains.

24. Lysenko was the infamous head of biology in the Soviet Union in the 1930s and 40s who resisted Mendel's discoveries concerning genetics because they did not conform with Marxist ideology purporting environmental determinism. With Stalin's backing, Lysenko forcibly suppressed alternative points of view.

25. Fuller claims that, "done properly," the philosophy of science is the application of political philosophy to scientists, a group with special status (1988, 6). Also recall his claim that sociological considerations should animate epistemology, which I cited earlier.

26. I say "mind-independent fact" to allow for the influence of beliefs or attitudes on certain facts. If I believe that I am depressed, for example, that may affect my actually being depressed by contributing to a pessimistic outlook.

27. Other bases for rational evaluation of an end would be its feasibility for an individual given that individual's particular abilities, knowledge, resources, and the like. For example, "You shouldn't apply for that position *now*, before you've acquired the requisite experience and skills."

28. For more extensive analysis of the rationality of ends, see Smith (2000), especially 46–51.

29. This is elaborated in far greater depth in Smith (2000), Rand (1964), and Peikoff (1991, especially chap. 7).

30. Gaut and Foot clearly recognize spiritual values, though they do not use that term. See Gaut (1997, 185), where he cites needs for meaningful work and close personal relationships; and Foot (2001, 15), where she cites the value of memory and concentration.

31. While we sometimes refer to the good of inanimate objects, note that this is intelligible only relative to the ends of living organisms. What is good for the car, for instance, is actually good for its owner or operator.

32. To differing degrees, the authors I referred to earlier as naturalists all embrace aspects of this view. Hursthouse, for instance, maintains that we do well to start in ethics by thinking about plants (Hursthouse 1999, 196). Gaut writes that "trees can have *good* roots because trees have goals, specified by their nature, and good roots are those which help achieve these goals" (Gaut 1997, 185, emphasis added). For an extended defense of the intimate relationship between goals and life, see Binswanger 1990. Wallace argues that an objective basis for normative theses, including the explanation of why certain traits are virtues, is found in life (Wallace 1978, 18). And in the most wholehearted recent attempt to provide naturalistic foundations for value Foot claims that the grounding of a moral argument is ultimately in facts about human life (Foot 2001, 24). Which traits are virtuous depends on facts about human needs, interests, and desires, just as claims about what is a good elephant depends on facts about elephants' needs, interests, desires (230). "For human beings the teaching and following of morality is something necessary. We can't get on without it" (16–17).

33. For further discussion of this, see Foot (2001, 15, 45–46).

34. While the roots of this account of value are biological, its aim is not sustenance simply of one's body but of one's life. For human beings, life encompasses far more than one's physical state. Correspondingly, *survival* refers not merely to the clinical clinging to that legal status but also to a flourishing life, the condition of an organism in peak condition, fit and thus poised to continue to exist. Gaut

offers a brief defense of this (Gaut 1997, 185–86). For a full explanation along with a discussion of the many questions it naturally raises, see Smith (2000, chap. 5), which is devoted entirely to the relationship between survival and flourishing.

35. For some further qualifications on the precise character of objective value, see Peikoff (1991, 241–42) and Smith (2000, 97, 120–21 n.30).

36. I thank Wolfgang Spohn for prodding me to clarify this.

37. Peter Railton (1986, 10) likens goodness to nutrition. All organisms require some nutrition, but not the exact same things.

REFERENCES

Binswanger, Harry. 1990. *The biological basis of teleological concepts.* Los Angeles: Ayn Rand Institute Press.

Damasio, Antonio. 1994. *Descartes' error: Emotion, reason, and the human brain.* New York: Avon Books.

Feyerabend, Paul. 1978. *Against method: Outline of an anarchistic theory of knowledge.* London: Verso.

Foot, Philippa. 2001. *Natural goodness.* Oxford: Clarendon Press.

Fuller, Steve. 1988. *Social epistemology.* Bloomington: Indiana University Press.

Gaut, Berys. 1997. The structure of practical reason. In *Ethics and practical reason,* ed. Cullity and Gaut, 161–88. Oxford: Clarendon Press.

Gergen, Kenneth. 1988. Feminist critique of science and the challenge of social epistemology. In *Feminist thought and the structure of knowledge,* ed. Mary M. Gergen. 27–48. New York: New York University Press.

Gould, Stephen Jay. 1981. *The mismeasure of man.* New York: W.W. Norton.

Grene, Marjorie. 1985. Perception, interpretation, and the sciences: Toward a new philosophy of science. In *Evolution at a crossroads,* ed. David J. Depew and Bruce H. Weber, 1–20. Cambridge, MA: MIT Press.

Haack, Susan. 1998. *Manifesto of a passionate moderate.* Chicago, IL: University of Chicago Press.

Harding, Sandra. 1991. *Whose science? Whose knowledge?* Ithaca, NY: Cornell University Press.

Hull, David. 1988. *Science as a process.* Chicago, IL: University of Chicago Press.

Hursthouse, Rosalind. 1999. *On virtue ethics.* New York: Oxford University Press.

Kuhn, Thomas S. 1962. *The structure of scientific revolutions.* Chicago, IL: University of Chicago Press.

Laudan, Larry. 1984. *Science and values: The aims of science and their role in scientific debate.* Los Angeles: University of California Press.

Leplin, Jarrett, ed. 1984. *Scientific realism.* Los Angeles: University of California Press.

Longino, Helen. 1990. *Science as social knowledge.* Princeton, NJ: Princeton University Press.

———. 2002. *The fate of knowledge.* Princeton, NJ: Princeton University Press.

Nuland, Sherwin B. 1988. *Doctors: The biography of medicine.* New York: Vintage Books.
Peikoff, Leonard. 1991. *Objectivism: The philosophy of Ayn Rand.* New York: Dutton.
Quine, W. V. 1953. *From a logical point of view.* Cambridge, MA: Harvard University Press.
——. 1960. *Word and object.* Cambridge, MA: MIT Press.
——. 1969. *Ontological relativity and other essays.* New York: Columbia University Press.
Railton, Peter. 1986. Facts and values. *Philosophical Topics* 14 (Fall):5–31.
Rand, Ayn. 1961. *For the new intellectual.* New York: Signet.
——. 1964. The objectivist ethics. In *The virtue of selfishness*, 13–39. New York: Penguin.
——. 1967. What is capitalism? In *Capitalism: The unknown ideal*, 11–34. New York: Penguin.
——. 1982. Causality versus duty. In *Philosophy: Who needs it*, 115–23. New York: Bobbs-Merrill.
——. 1988. Who is the final authority in ethics? In *The voice of reason*, ed. Leonard Peikoff, 17–22. New York: New American Library.
——. 1990. *Introduction to objectivist epistemology.* 2d ed. New York: Meridian.
Smith, Tara. 2000. *Viable values: A study of life as the root and reward of morality.* Lanham, MD: Rowman & Littlefield.
Wallace, James D. 1978. *Virtues and vices.* Ithaca, NY: Cornell University Press.

9

On the Objectivity of Facts, Beliefs, and Values

Wolfgang Spohn
Department of Philosophy, University of Konstanz

Rather than attempting to advance the well-belabored topic of objectivity, my focus in this chapter is modest: to introduce some distinctions for conceptual clarity, to sketch the routes to objectivity that appear most plausible against the background of the current discussion, and to point out some errors that unnecessarily heighten present confusions.

I distinguish between a notion of objectivity accruing to facts and thus to objects and properties and building on ontological independence and a quite different notion of objectivity accruing to beliefs and idealizing intersubjective agreement. I will show that the latter notion is in principle applicable to values as well. The difference is only that there are stronger specific reasons for hoping for objectivity in the case of beliefs and stronger specific reasons for restricting claims to objectivity in the case of values. I begin with the objectivity of facts and then proceed to that of beliefs and values.

On the Objectivity of Facts

When Hume was struggling with and despairing of the objectivity of objects, his concern was about the continuity and independence of external objects.[1] External to what? The answer is not really important; I shall return to the point. Continuity is not so important either; there are also discontinuous but perfectly objective objects.[2] The crucial feature of objects that makes for their objectivity is their independence. How is independence here to be understood?[3] It is *ontological independence from us*; this has always been intended as one basic sense of objectivity.

What is ontological independence? An object X ontologically depends on an object Y if X cannot exist without Y, if it is metaphysically impossible that X exists, but not Y. For instance, each human being ontologically depends on its mother, or has its mother essentially. Conversely, X is ontologically independent from Y, if X might exist even if Y does not exist.

And who is "us" in this context? There is a little vagueness here. If it is to be the whole of humankind, then Henry Ford or, respectively, my ancestors and I must be exempt from it in order for Ford or myself to acquire objective existence. So, "us" rather refers to human minds. This secures my objective existence; I may lose my mind and still exist. But the vagueness persists. My perceptions ontologically depend on my mind but not on other minds or objects[4]; still, I do not see why my perceptions should be denied objective existence.

The same vagueness affects, by the way, the notion of an external object. *External* means here "spatially external to us," and so we might again wonder who is us. But it is not so important to resolve the vagueness. The objectivity issue is not about potential borderline cases like my perceptions, me, or my Ford, but rather about objects in general, no matter how we draw the borderline.

In one way or the other, it was always skepticism that motivated the general denial of the objective existence of objects. If we follow, for instance, Berkeley's idealistic doctrine *esse est percipi* (to be is to be perceived) or conceive of external objects in a phenomenalistic way as mere constructions from sense data, then objects ontologically depend on perceptions or perceiving subjects and are not objective.[5] Or if we follow more sophisticated Kantian idealism or its modern heritage, the many variants of constructivism, then the external world and its objects ontologically depend on our concepts; the world and the objects could not be what they are if our concepts would not be what they are, if we would conceive them with different concepts. Again, they are not objective.[6]

I think, and many have thought so before, that this comes to a straightforward refutation of all these kinds of phenomenalism or constructivism. No motive can be good enough for going ontologically crazy. We may indeed be tempted to endorse such doctrines, but then the only reasonable response is to stop the temptation and to inquire what has led us so far astray.

This might have been the skeptical question — how do we know that objects are objective? — temptingly suggesting that we cannot really know. Well, how do we know anything? And in particular how do we know about counterfactuals (such as objectivity claims)? These are good questions, and even if we have some way of answering the first general one, the second about counterfactuals may remain mind-boggling. There is every reason for the vigorous interest in counterfactuals in the last thirty-five or fifty-five years. Still, with these questions we are leaving ontology and entering epistemology, and there is no ontological conclusion to be drawn from epistemological issues and in particular no way to resort from epistemological despair to ontological insanity.

Another thought leading us astray is that it is *we* who cut up the world into parts and pieces, who constitute objects, that we do so with our conceptual power, and that we might do it differently if we faced the world with different concepts. This sounds true, and it seems to directly lead to some kind of constructivism. I think it *is* true, but it has no such consequences. It is simply false in general to say that an object would not exist unless constituted (by us or whomever) or that there would be different objects if we constituted them differently. This may apply to some special objects, for instance, our concepts themselves. But in general there is no way for the constructivist to get beyond such counterfactual trivialities as: if nobody conceived or named objects, nobody would conceive or name them, or if we would constitute different objects, we would think and talk about different objects, or without constitution no constitution. Any stronger constructivist claim rests on a misunderstanding of constitution. The objects do not come to exist by getting constituted. The whole point of constituting objects is normally to fix their constitutive or essential properties in such a way that the constituting or fixing does *not* belong to them; hence, they would exist just as well unconstituted. Constituting objects and ruling how to correctly talk about them counterfactually are one and the same matter, and we do it usually so as to acknowledge the ontological independence of objects from us, that is, their objectivity.

The same holds, *mutatis mutandis,* for properties. The talk of existence and ontological independence refers only to objects, not to properties. Still, we may say that a property is *objective* insofar as its extension, whether or not it applies to a certain object, is counterfactually independent of us or our minds, of what we or our minds do.

Colors provide a striking example. The relation "X looks red to Y" is not objective in this sense. That a certain tomato looks red to me counterfactually depends on my existence, on my being awake, on my not wearing strongly colored glasses, and so on (though it does not counterfactually depend on the existence of others — recall the vagueness of "us"). The same holds for the property of being appeared to redly, as Chisholm once phrased it. The contrary, however, holds for the property of *being red*. This tomato is red, and it would still be red, even if it is unperceived or if there is nobody at all or only organisms with a different color vision to perceive it. These counterfactual claims seem obviously true to me. They are also compatible with the fact that which property is actually denoted by "being red" depends on our neurophysiological and mental make-up (in ways we understand still quite incompletely). But this dependence cannot be expressed by counterfactuals about what is or would be red. The property of being red itself is no longer dependent in this way.[7] Hence, being red is an objective property, and so are other properties all the more.

Objects and properties make up states of affairs, and, if they obtain, facts. This is why the title of this section summarizes the observation made so far.

I shall, however, not deepen now these observations and pursue this notion of objectivity any further. One obvious reason is that there would be no end in doing so. A more specific reason is that I would like to discuss values in the end, and that this notion of objectivity does not seem to be helpful when it comes to values. In any case, I do not see how values could be objective in this sense. Values seem to ontologically depend on valuers. Things can be red in a possible world without perceivers, but things cannot be valuable in a possible world without valuers. One may argue about who is a valuer; presumably not only humans. But stars and stones are not, I take it, and hence nothing has any worth in a world in which there are only stars and stones. They may do this or do that or all disappear in a black hole; nobody cares.[8] Hence, if this is right, we have to look for other notions of objectivity in order to possibly make sense of objective values.

Still, the present ontological notion of objectivity is perhaps the most basic one deeply worrying, as we have seen, philosophy for centuries. And it is well worth being reminded here, because the deeper philosophy of science gets involved in the epistemological workings, the more careless it gets about the ontological dimension, and the more easily it slips into talking nonsense. Let me add, therefore, just one general remark: We are always prone to mix ontological and epistemological issues; and we often hardly notice it. The best measure against this danger is to move within a framework where the two dimensions are clearly separated from the outset. Such a framework exists and is even formally codified. It is called two-dimensional semantics nowadays, but this title hides its origins. It has also been called, circumstantially, the epistemological reinterpretation of the character theory, where "character theory" refers to David Kaplan (1977) and "epistemological interpretation" to Robert Stalnaker (1978) and both refer further back to the ground-breaking attempts of Saul Kripke (1972) and Hilary Putnam (1975) to disentangle metaphysical and epistemic necessity. As the name says, these two dimensions *and* their relation are formally represented in two-dimensional semantics.[9] Of course, the details of interpretation present difficult issues (and sometimes perhaps old issues in a new disguise).[10] Still, the framework is the best device we have for sorting out ontological-epistemological confusions.

For instance, to mention a particularly relevant example, this framework makes clear that the notion of intension, which was once to explicate the notion of meaning in the sense of truth condition, is systematically ambiguous. There are diagonals and intensions (Stalnaker), primary and secondary intensions (Chalmers), or A- and C-intensions (Jackson). Armed with this distinction one can easily see that the big debate between empiricists (Carnap and others) and their relativistic outgrowth (Feyerabend, Kuhn) on the one hand and realists (Putnam, Boyd, and so on) on the other simply falls victim to that ambiguity. The two parties are talking at cross-purposes about different kinds of intensions or meanings, and both of them may be accommodated in the two-dimensional framework (see Hass-Spohn 1997, sec. 4).

On the Objectivity of Beliefs

Let us move then to the other important notion of objectivity that relates to beliefs rather than to facts. What are we seeking here? The point is not about the having of beliefs. Whether or not I have a given belief is presumably just as objective as any other fact. It is rather the objectivity of belief *contents* we are interested in. This remark gives rise immediately to the observation that our inquiry of objectivity is about to cover the two-dimensional scheme just alluded to. We have talked about the objectivity of facts, or intensions or secondary or C-intensions, and we now attend to the objectivity of belief contents, or diagonals or primary or A-intensions.

What are we after, though, with the objectivity of belief contents? Clearly, the issue is about the objective truth, or the objective falsity, of belief contents. If we identify, however, belief contents with states of affairs or somehow treat them on a par and thus reduce the truth of a belief content to the obtaining of a state of affairs or a fact,[11] then we are back at the issue of the previous section. Hence, we should look for a different approach. There is indeed one; we all have the idea that in principle everybody should be able to agree to objective truth and to reject objective falsity, and that there is no objectivity to the extent that there is no such agreement. The idea at least sounds different from the one discussed in the previous section. So, let us pursue this a little bit and then see whether it is really different.

Let me start by defining that an assertion or a statement p (or rather its content, because this, and not the linguistic form, is what we are interested in) is *shared* by a group G or *intersubjectively accepted* in G if all members of G who have a judgment on it accept or endorse p; and we may call p *intersubjective* in G if acceptance of p or rejection of p (or suspension of judgment concerning p) is shared in G, and *subjective* in G otherwise. These relative notions are introduced only for the sake of explicitness. Usually, some large group tacitly understood is the standard by which subjectivity and intersubjectivity are measured, say, the linguistic community of all Americans or scientific communities sharing some expert language.

For instance, "I am tired" is clearly subjective, because I am accepting it as I am writing these words. "Philosophy is fun" is shared, I hope, by all readers of this book (at least before reading it), but it is clearly a minority opinion in larger groups. However, "philosophy is easy" is, I take it, intersubjective, since everybody thinking about it would deny it.

Now, the crucial question is, of course, what has intersubjective acceptance to do with objective truth? Not much, it seems. Each person is fallible in principle (though there may be a few things on which each is infallible); this is a lesson we have learned for a long time. Hence, subjective acceptance and truth may (almost) always fall apart.

The same holds, then, for any actual group of persons. Why should a million people differ in principle from a single person? Some bit of counterfactuality does not change much here. If all the people in the relevant group would get all the information they can actually get, would be perfectly rational, would freely discuss their issues, and would then agree on a critical and considered judgment, they still might be wrong. The sources of error need not lie in factual imperfections of the group. All the evidence actually available to us might be objectively misleading. Hence, any crude or mildly refined consensus theory of truth[12] will not do.[13]

This is not to say that there is no relationship at all between intersubjective acceptance and objective truth. A lot of counterfactuality *does* change the picture. The Peircean limit of inquiry always appeared to me to be an attractive idea. The consensus reached after all *possible* experience is acquired and all powers of reasoning are perfectly exerted must be true. Note that the consensus is no longer doing any real work here. Counterfactually, everybody has the same complete evidence including information about what the others think and the same ideal powers of reasoning. And then, we might suppose, consensus falls in.

The ideal is indeed vastly counterfactual. We cannot travel far enough in space-time in order to inquire what is going on at every space-time point; our memory and our computation are much too poor to process the counterfactual evidence; and we cannot undo the effects of our investigation and let the world go on as if unobserved. But under this vastly counterfactual assumption all possibility of error is excluded; nothing can turn up to prove us wrong. Or, as Putnam (1981) would have it, the ideal theory must be true.[14]

Is this really so? There are two opposite lines of attack, one saying that Putnam's claim is wrong and the other that it is too trivially true to be of any value. Let me briefly discuss the two criticisms.

The metaphysical realist insists that the ideal theory might still be wrong, an assertion Putnam fought with so many words. However, we know by now, indeed through Putnam's help, that the "might" might indicate metaphysical or epistemic possibility, and then the metaphysical and the internal realist do not seem to have any dispute. Under the metaphysical reading the metaphysical realist is clearly right. If we have acquired complete evidence about the actual world and build our ideal theory on that basis by exerting our ideal powers of reasoning, if we have thus reached the ideal limit of inquiry concerning the actual world, there is still a possible world in which we have acquired the same evidence and build the same ideal theory, but in which this theory is wrong. I do not find it particularly difficult to imagine such a possible world; one simple trick is that the evidence is supposed to be complete only with respect to the actual world but not with respect to the imaginary world. Hence, it cannot be this metaphysical possibility that Putnam is fighting. It is rather an epistemic impossibility that the ideal theory is false. This theory cannot *turn out* to be false; there is nothing left that could show us that we erred. This appears obvious, but the metaphysical realist need not take exception to it.

What about the other criticism that it is too obvious? What is the criticism here? It is that the ideal limit of inquiry, the ideal theory, and the like are empty notions (see, for example, Brandom 1994b). What we would believe in this counterfactual limit does not inform us about objective truth; rather, we have to appeal to objective truth in order to arrive at a correct description of those counterfactual conditions excluding error. If we thus have to build truth into the ideal, it is not a deep philosophical discovery that the ideal comes out true.

I do not agree with this criticism. I believe instead that internal realism, the claim that the ideal theory cannot turn out to be false, has substantial consequences concerning the relationship between rationality and truth. To this extent, I am a good Kantian. Indeed the consequences cut two ways.

One way is that the accessibility of truth constrains the rationality of belief states. For instance, the so-called Reichenbach axiom in inductive logic postulating that subjective probabilities must converge to observed relative frequencies is an axiom in this spirit and it puts a severe constraint on a priori subjective probabilities (see, for example, Carnap 1980, sec. 20). Extensive and systematic investigation about such constraints are provided by formal learning theory (see, in particular, Kelly 1996).

The other way is that the rationality of belief states constrains truth. Whatever the state is that is reached in the ideal limit, it must still be governed by the laws of rationality, and to the extent we can establish such laws without recourse to truth we may thus draw conclusions concerning the structure of truth. For instance, the probability axioms, or the axioms of AGM (Alchourrón-Gärdenfors-Makinson) belief revision theory (see, for example, Rott 2001), or the axioms of ranking theory (see, for example, Spohn 1988), provide such laws, which we can seek to strengthen. Thus we may arrive at principles a priori true in all possible limits of inquiry.[15] Indeed, I guess most of the attempts to transcend correspondence, semantic, or deflationary views of truth can be, perhaps somewhat swiftly, subsumed here; I think, for example, of pragmatic truth theories, the coherence theory of truth (as elaborated by Rescher 1973), evaluative theories of truth (see Ellis 1990, chap. 7, in particular), or even Brandom's (1994a) expressivistic and inferentialistic strategy.[16] All these approaches have at least the potential of arriving at substantial a priori truths.

This provokes a side remark: Philosophers still struggle with disentangling a priori truth and analytic truth. This is difficult because both seem to be conceptual truths. Kripke made a distinction by defining analyticity as an a priori necessity. This was revolutionary since it opened the non-empty category of a priori contingencies. But the conceptual was still the only source of the a priori. The previous paragraph points the way to a broader base of the a priori consisting of the structure of reason as a whole, and there is more to reason than concepts.

In any case, the rebuttal of the objection that the ideal limit of inquiry presupposes the notion of objective truth is important in our present dialectic situation. If we would have been forced to grant the objection, then we would have to admit as well that what we are heading for presently is not really a new notion of objectivity. Rather, we would have to start with objective truth, the limit of inquiry would approach just that and would not add an independent element, and then it is difficult to see in which ways the objectivity of truth differs from the objectivity of facts already discussed. But, if I am right, this is not our worry, and we may assume to have captured a different notion of objectivity, namely as intersubjectivity in that vastly counterfactual community reaching the ideal limit of inquiry.

How far does objectivity in this sense extend? Not nowhere, not everywhere; it depends. And it may not be easy to answer; after all, we are speculating about what would happen in that vastly counterfactual limit. Still, there seem to be many clear cases.

For instance, whether there was a dinosaur standing exactly at the place where I am sitting now exactly twenty-five billion days ago is an objective matter. In the ideal limit we all would agree on it even if we actually have no ways at all to find out about it. This seems to hold for all reports of singular facts about any place and time in the world. Moreover, we say that the theories of present-day science are better than those of former centuries, not simply different ones. This shows at least that we are confident in principle to be able to convince everybody including our colleagues from the former centuries of the theoretical progress and that we thus treat the theories as objective. I do not see why this should be unjustified.

What I would like to resist, in particular, is the thought that we are barred from objectivity and are at best struck with intersubjectivity within our linguistic community simply because we all are prisoners of our conceptual schemes. Conceptual schemes can be changed, indeed reasonably changed, they can be translated, and they can be merged, indeed translation may rather be merging, and so on. These processes are difficult to describe, and there is a lot of philosophical dispute about them. But, no doubt, they are effective. Hence, I do not see why we should be bound to end up with several incommensurable conceptual schemes. On the contrary, empirical concepts essentially refer to objects and properties that are objective in the first sense and thus not constituted by these concepts, and this at least opens the way for translating and unifying conceptual schemes.

There are also many cases of subjective assertions. Indexicality is one clear source of nonobjectivity. "I am tired" is something I accept now, but not always, and many do not accept it. Sometimes, the relativity of an assertion to the circumstances of its utterance is written into the semantic rules of its expressions as in the case of indexicals. Sometimes, it is more implicit: "Lucas is tall?" It depends on the context. "Philosophy is fun?" For me, but not for everybody. Hence, all such in one or the other way relative assertions are not objective. However, they can be turned into objective ones by having their relativization made explicit. "Wolfgang is tired on January 6, 2002, at 1 a.m.," "Lucas is tall for his age, but not within his class," "philosophy is fun for Wolfgang": all this is perfectly objective. If the source of subjectivity is relativity, then, it seems, objectivity can always be reached by appropriate explicit relativization.[17]

There is a further clear case of nonobjectivity in the present sense, namely the case of self-verifying beliefs or, more generally, of beliefs causally affecting their own truth. If we believe the economic future is bright, it is bright, but not if we believe it is dull. The medicine helps if I believe it helps; it does not if I do not believe in it. If I expect my kids to be well-behaved, they are not; if I fear my kids to be naughty, they are not, either. These are strange ways of positive or negative feedback. And they hamper objectivity. The point is not relativity of content. We can eliminate the relativity and still have the feedback. The point is that what is accepted in the limit causally depends on what is accepted in the beginning, and hence there is no unique limit and thus no objectivity. It was generally sensed that there is something fishy about these self-confirming or self-affecting beliefs. It consists precisely in their deviant lack of objectivity; they do not behave like normal beliefs.

I have been careful at the beginning of this section to generally define the subjective and intersubjective acceptance and the objective acceptability of *assertions*. In my examples I have considered only assertions expressing beliefs (except perhaps "philosophy is fun"). Hence, I was so far only discussing the objectivity of beliefs. But the definitions clearly apply as well to other kinds of assertions that express other kinds of attitudes. Thus we can discuss the objectivity of these further kinds of assertions on the same model as those of beliefs.

On the Objectivity of Values

The paradigm assertions for which objectivity issues have become historically eminently important are, of course, religious ones.[18] Religious assertions are partly descriptive; they contain a picture of how the world is and works. But this picture does not exhaust their content. Rather, a bunch of different attitudes, in particular moral and evaluative ones, indeed a whole form of life, meshes in them. I cannot see why their descriptive content as such should not be objective.[19] But their content as a whole is certainly not objective, as a long and bloody history has taught us; at best intersubjective agreement within religious communities can be reached. The insight, though, that religious assertions are *not* objective, is, or is generally thought to be, the axe at the root of religion. This is why religious tolerance is so hard to reach.

The example shows the enormous difficulties arising from mixed discourse. If one sides with Tara Smith (2000) on the objectivity of evaluative discourse, then it mixes well with objective descriptive discourse.[20] If not, one faces the task of disentangling the mixed discourse into components of varying objectivity, often a difficult task, as we just saw, and one we shall not pursue now. Let us rather only try to get clearer on one particularly important component, the objectivity of evaluative or normative assertions.[21]

The central claim of Smith (2000, chap. 4) is that life is the source of all values. Perhaps the source is smaller; I doubt that flowers or earthworms are capable of values. But, certainly, inanimate objects cannot have values. Smith continues that at the same time life is itself a value or an end; indeed, it is "the ultimate value insofar as all other genuine values depend on it and are the means to it" (104). It is an innate value of living beings to maintain life,[22] including flourishing and not mere vegetating. And all the values that we objectively have according to Smith (2000) derive from this fundamental value.

However, I do not think that the derivation can succeed. One reason for my doubt stems from a general analogy. Willard V. O. Quine repeatedly claimed that truth is the ultimate value of all cognitive activity and that normative epistemology is just a branch of engineering, the technology of truth-seeking (see, for example, Hahn and Schilpp 1986, 664ff.).[23] Thus, epistemology becomes naturalized. In view of the utterly rich normative discussion in epistemology this claim always appeared to me to be entirely unredeemed. Perhaps epistemology can be unified under this goal, but nobody can claim to have shown this.[24] And what is true of epistemic norms is all the more true of norms and values in general.

This has only been an analogy so far, not an argument. So let us look directly at values. Some values seem to belong to our biological endowment and thus to be objective. For instance, "I am alive" or "I am without pain" are assertions the content of which everybody values. No doubt, but this is due to the indexicality of these assertions. Making their relativization explicit has here a *subjectivizing* effect, opposite to the case of beliefs. "Wolfgang is alive" and "Wolfgang is without pain" are valued by me but not necessarily by my neighbors. Hence, the values we are biologically endowed with are, on the contrary, highly subjective.[25]

This fact is at the root of all moral conflicts and of all difficulties of attaining evaluational objectivity. Usually, staying alive is a cooperative enterprise to the benefit of all, but much too often we have to weigh life against life (and even life against money). Often, flourishing is a mutual activity, and often flourishing is exclusive and excessive at the expense of others. Smith (2000) is, of course, aware of such problems and discusses them in chapter 6 of that book. I do not think they dissolve so easily.

Christoph Fehige (2003) tries to stop the argumentative gap right at the beginning. He argues that I should value not only my absence of pain or, more generally, my positive hedonic states but also those of all others. Empathy is an a priori moral principle, he argues. This is at least plausible. If so, however, it follows that my positive satisfactive states should be valued, and the negative ones disvalued, by everyone, not only by me. This does not yet quite mean that these values are objective ones. Fehige rather tries to establish even more, namely that my hedonic states should be valued by others *to the same degree* as I do, and vice versa. Indeed, only then would the objectivity of the evaluation of the hedonic states, which is a quantitative, not a qualitative, affair, be guaranteed.

The argument is an intricate one, but I doubt whether it succeeds. To this extent it is doubtful whether we have reached an objective base of our evaluations. But even if we would have reached it, we could build far less on it than, for example, Smith (2000) thinks. In arguing this, I have a little terminological difficulty. The talk of hedonic or satisfactive states is mine, not Smith's. She is not a hedonist. Her concern is rather living and flourishing, which she takes in an Aristotelian sense. And given the objective values of living and flourishing, she argues that it is an objective empirical question when and how we stay alive and when and how we flourish. Hence, the conditions under which we stay alive and flourish are objectively valuable as well. However, flourishing has certainly a lot to do with hedonic states. And the hedonist might take a similar stance, namely that it is an objective empirical issue when and how our hedonic states are pleasant and not painful and that further values may thus be objectively inferred from the value of our hedonic states. It is the latter claim I want to oppose, but my point applies, it seems, just as well against Smith's stance.

My point is simply that the self-confirming structure that we found to be rare and deviant in the case of beliefs is quite common in the case of values. Or more generally, it is not at all rare or deviant but perfectly normal that values causally affect their own acceptability and maintainability. If I desire to be a philosopher, if being a philosopher is a subjective value for me, being a philosopher makes me happy; but it would be agonizing to be a philosopher without valuing it. If a pregnant woman wants to have a baby, she will derive immense satisfaction from it. If she rejects it, she is likely to be torn in conflict and ambiguity. These are normal self-confirming cases. But the structure may be more pathological. A desire may be so insatiable that each fulfillment of the desire is more painful than pleasant. A man may long for inaccessible women; as soon as a woman responds to his courting, he starts rejecting her. The list of examples is endless.[26]

The general logical point of these examples is this: Very often, or even usually, so-and-so much valued satisfactive states are causally affected not merely by the realization of a valued state of affairs but also by that state of affairs being valued so-and-so much. Hence, the value of that state of affairs cannot be simply derived from its propensity to produce happiness. Rather, its value must somehow be given beforehand, and then its evaluation can be confirmed, or disconfirmed, by its propensity to produce happiness as one valued to this degree. So, what we humans have to seek is rather a kind of equilibrium between our evaluations of the various states of affairs and their causal effect on hedonic states, if thus valued. There are as many equilibria as there are happy persons or flourishing and fulfilled lives. Each one is subjective, and so are the values maintaining it.[27]

The point just sketched reappears, I believe, on the intersubjective level. The constitution of societies — understood widely as the collection of their legal codes, moral rules, social values, customs, and so on — may vary greatly, without there being an external arbiter proving them wrong or immoral or unacceptable. In the positive case, various religious communities may be happy and stable without suppressing their members. The balance between individual rights and social duties, between liberalism and communitarianism may take many forms. All societies struggle with intergenerational justice, and they find many different solutions. And if our societies are discussing such problems and changing their terms, they need not be moving toward an objective ideal but rather only toward an improved internal equilibrium.

This is not a pleading against patronizing or for saying *de gustibus non est disputandum* (there is no accounting for tastes), which is only very limitedly true. It is only a reminder that one should be very clear about one's base for criticizing and patronizing others. Often, this base is not objective, but subjective. Either in the sense that one declares one's own subjective point of view; this tends to be futile. Or in the sense that one criticizes the other's subjective point of view for its *internal* inconsistencies, where "inconsistency" takes on a broader range in light of the preceding paragraph. For instance, a man's longing for inaccessible women is broadly inconsistent in the relevant sense; one should help, that is, patronize him, and not let him remain in his unhappy condition.[28]

All this is even not a pleading against the objectivity of values, against an objective basis for such criticism. I hope and believe that there is such objectivity. Indeed, most do; at least, most moral issues are discussed as if there would be an objective answer. One must grant, of course, that this objectivity is in no way settled. This is what moral philosophy is all about. There are rival conceptions of moral norms and values, there are rival justifications for them, and the claims of objectivity depend on the quality of these justifications. If, for instance, the long sought derivation of substantial moral principles from principles of rationality alone finally succeeds, the two kinds of principles are distinguished by the same degree of objectivity. Whether the justifications are compelling enough to produce general agreement, whether the limit of moral inquiry is unique and the unity of practical reason perfect, is nice to speculate but hard to argue about.

But even if the answer is positive, the extent of the objectivity of norms and values is open. Basic moral principles like the Golden Rule, the Categorical Imperative, or the Principle of Universalizability are good candidates; basic human rights are so as well. But where the boundary is to be drawn above which objectivity reigns and below which the various reasons for nonobjectivity specified above set in, that is hard to tell.

In any case, though, the objectivity in normative and evaluative matters differs from that in descriptive matters. Normative or evaluative discourse is autonomous, not reducible to descriptive discourse about empirical matters. And unlike the latter, the objectivity to which it possibly raises is not helped by the ontological independence and the objectivity of facts, properties, and objects.

NOTES

1. See the section "Of Scepticism with regard to the senses" in Hume 1739/1740, I, iv, 2; see also, for example, Bennett (1971, chap. 13).

2. The journal *Erkenntnis*, for instance, which existed from 1930 to 1939 and then from 1975 to the present, definitely in an objective manner, as I am privileged to say.

3. Independence is one of the worst notions in philosophy, equally multiply ambiguous as, say, necessity.

4. At least, this is what one would normally say. But there may be sophisticated individuations, such as externalistic accounts of perceptions and perceptual contents according to which their ontological dependence spreads further.

5. If the perceiving subject is God and if God is objective, the case is different, of course.

6. See also the devastating and, I think, justified criticism of the various kinds of constructivism in Devitt (1984, chaps. 8–14).

7. This may sound like objectivism about color according to Jackson and Pargetter (1987). But in fact, it is my own, I think, more defensible version explained in detail in Spohn (1997).

8. See also the related criticism of Zangwill (2001).

9. Best elaborated perhaps in Chalmers (1996), Haas-Spohn (1995), and Jackson (1998), but see also Kupffer (2003).

10. Clear differences already show up in the works referred to in the preceding note.

11. The reader will notice that I am somewhat clumsily avoiding the notion of a proposition — precisely because it has played an unfortunate double role like the notion of intension, which is clearly split up in two in two-dimensional semantics.

12. Such as the one of Habermas (1973), who does not refer to stronger counterfactual conditions than the ones just mentioned; however, he claims only to state a criterion of truth, whatever this is.

13. As Tara Smith has rightly observed in her contribution to this volume, from comments on which my contribution has emerged.

14. Putnam says, "Truth is an *idealization* of rational acceptability" (1981, 55). "The two key ideas of the idealization theory of truth are (1) that truth is independent of justification here and now, but not independent of *all* justification. . . . (2) truth is expected to be stable or 'convergent'" (1981, 56).

15. In Spohn (1991, 1999) I tried to establish some weak coherence and causality principles in this way.

16. Though Brandom there defends a deflationary theory of truth (sec. 5, V), the further reaching goal I am attributing to him returns under the heading "objectivity."

17. See also Mühlhölzer (1988), who elaborates and defends an explicitness conception of objectivity the clearest expression of which he ascribes to Hermann Weyl.

18. This section shows most clearly the origin of this chapter as a commentary on Tara Smith's contribution to this volume. She has very pronounced opinions on the objectivity of values that are also part of her present contribution. I agree with many things she says, but not this part.

19. Indeed, it is objectively false, I assume; I cannot believe that any of the religious world pictures survives in the limit of inquiry.

20. As she confirms in Smith (2000, 101ff).

21. Perhaps I am a bit careless in this section in interchangeably talking of norms and values. But whatever the distinctions to be drawn here, they do not become relevant in our present context.

22. Until its natural end — where different ideas about the natural end of life seem to be encoded in different species.

23. This is just the uninteresting sense, discussed above, of the ideal limit of inquiry reaching the truth.

24. Indeed, I interpret Quine (1995, 49–50) as stepping back a bit from his earlier bold claims.

25. One needs to be careful here. "Wolfgang values absence of pain" describes an objective fact (in the first sense) and expresses an objectively true belief (in the second sense), but "Wolfgang is without pain" is subjectively valued by me, not necessarily intersubjectively by everyone.

26. Elster (1979, 1983) is famous for collecting and analyzing many of such more-or-less paradoxical situations. One must pay attention, though. There is great variation in the logical structure and analysis of these situations.

27. This description may sound confusing, or confused. In Kusser and Spohn (1992) we analyzed the problem (which we traced back at least to Joseph Butler) in detail and with decision-theoretic precision, and we tried to disentangle the confusions that invariably adhere to brief sketches of the problem like this one. We are convinced that the problem has important consequences for decision theory and for the view of practical rationality in general.

28. For a more systematic development of this thought see Kusser (1989); for a different approach see, for example, Brandt (1979).

REFERENCES

Bennett, J. 1971. *Locke, Berkeley, Hume*. Oxford: Oxford University Press.
Brandom, R. 1994a. *Making it explicit*. Cambridge, MA: Harvard University Press.
———. 1994b. Unsuccessful semantics. *Analysis* 54:175–78.
Brandt, R. 1979. *A theory of the good and the right*. Oxford: Clarendon Press.
Carnap, R. 1980. A basic system of inductive logic. Part 2. In *Studies in inductive logic and probability*, vol. 2, ed. R. C. Jeffrey, 7–155. Berkeley: University of California Press.
Chalmers, D. J. 1996. *The conscious mind*. Oxford: Oxford University Press.
Devitt, M. 1984. *Realism and truth*. Oxford: Blackwell.
Ellis, B. 1990. *Truth and objectivity*. Oxford: Blackwell.
Elster, J. 1979. *Ulysses and the sirens: Studies in rationality and irrationality*. Cambridge: Cambridge University Press
———. 1983. *Sour grapes: Studies in the subversion of rationality*. Cambridge: Cambridge University Press.
Fehige, C. 2003. *Soll ich?* Stuttgart: Reclam.
Haas-Spohn, U. 1995. *Versteckte Indexikalität und subjektive Bedeutung*. Berlin: Akademie-Verlag.
———. 1997. The context dependency of natural kind terms. In *Direct reference, indexicality, and propositional attitudes*, ed. W. Künne, A. Newen, and M. Anduschus, 333–49. Stanford, CA: CSLI Publications.
Habermas, J. 1973. Wahrheitstheorien. In *Wirklichkeit und Reflexion: Walter Schulz zum 60 Geburtstag*, ed. H. Fahrenbach, 211–65. Pfullingen: Neske.
Hahn, L. E., and P. A. Schilpp, eds. 1986. *The philosophy of W. V. Quine*. La Salle, IL: Open Court.

Hume, D. 1739/1740. *A treatise of human nature.*
Jackson, F. 1998. *From metaphysics to ethics.* Oxford: Clarendon Press.
Jackson, F., and R. Pargetter. 1987. An objectivist's guide to subjectivism about color. *Revue Internationale de Philosophie* 41:127–41.
Kaplan, D. 1977. Demonstratives. An essay on the semantics, logic, metaphysics, and epistemology of demonstratives and other indexicals. Manuscript. Published in *Themes from Kaplan,* ed. J. Almog, J. Perry, and H. Wettstein, 481–563. Oxford: Oxford University Press, 1989.
Kelly, K. T. 1996. *The logic of reliable inquiry.* Oxford: Oxford University Press.
Kripke, S. A. 1972. Naming and necessity. In *Semantics of natural language,* ed. D. Davidson and G. Harman, 253–355, 763–69. Dordrecht: Kluwer.
Kupffer, M. 2003. Kaplan's a priori. *Forschungsbericht, no. 104, der DFG-Forschergruppe "Logik in der Philosophie."* Universität Konstanz.
Kusser, A. 1989. *Dimensionen der Kritik von Wünschen.* Frankfurt: Athenäum.
Kusser, A., and W. Spohn 1992. The utility of pleasure is a pain for decision theory. *Journal of Philosophy* 89:10–29.
Mühlhölzer, F. 1988. On objectivity. *Erkenntnis* 28:185–230.
Putnam, H. 1975. The meaning of "meaning." In *Mind, language, and reality,* vol. 2 of *Philosophical papers,* ed. H. Putnam, 215–71. Cambridge: Cambridge University Press.
———. 1981. *Reason, truth, and history.* Cambridge: Cambridge University Press.
Quine, W. V. 1995. *From stimulus to science.* Cambridge, MA: Harvard University Press.
Rescher, N. 1973. *The coherence theory of truth.* Oxford: Oxford University Press.
Rott, H. 2001. *Change, choice, and inference.* Oxford: Oxford University Press.
Smith, T. 2000. *Viable values.* Lanham, MD: Rowman & Littlefield.
Spohn, W. 1988. Ordinal conditional functions: A dynamic theory of epistemic states. In *Causation in decision, belief change, and statistics,* vol. 2, ed. W. L. Harper and B. Skyrms, 105–34. Dordrecht: Kluwer.
———. 1991. A reason for explanation: Explanations provide stable reasons. In *Existence and explanation,* ed. W. Spohn, B. C. van Fraassen, and B. Skyrms, 165–96. Dordrecht: Kluwer.
———. 1997. The character of color predicates: A materialist view. In *Direct reference, indexicality and propositional attitudes,* ed. W. Künne, A. Newen, and M. Anduschus, 351–79. Stanford, CA: CSLI Publications.
———. 1999. Two coherence principles. *Erkenntnis* 50:155–75.
Stalnaker, R. C. 1978. Assertion. In *Syntax and semantics.* Vol. 9: *Pragmatics,* ed. P. Cole, 315–32. New York: Academic Press.
Zangwill, N. 2001. Against moral response-dependence. *Erkenntnis* 55:271–76.

10

A Case Study in Objectifying Values in Science

Mark A. Bedau
Reed College

Background to an Objectification of Values

There are at least two different ways to connect values and science. One is through the evaluation of science, and the other is through the scientific investigation of values. The evaluation of science is a nonscientific, political, or ethical investigation of the practices of science. Various proposed and actual scientific practices call out for social and ethical evaluation. A few that have received recent attention are the human genome project, intelligence testing, and encryption algorithms. Such evaluations of science contrasts sharply with what I call "the science of values." This is not one science or even one unified nexus of scientific activities but a loosely defined grab bag containing all scientific investigations of matters involving values.

One part of the science of values concerns what individuals or groups value or take an interest in; these are values considered from a first-person point of view. The values can concern anything, including morality, aesthetic matters, religion, politics, lifestyle, and livelihood. The science of first-person values includes such things as psychological studies of the values of individual people, sociological studies of the values of social groups, and anthropological comparisons of the values of different cultures.

Another part of the science of values concerns what is good for, or promotes the interests of, some individual or group from an external, third-person point of view. The subject whose interests are being studied might or might not internalize the values used in the external evaluation. Examples of the science of third-person values include studies of the value of a college education, of regular visits to the dentist, or of growing and eating organic food. They also include biological studies to determine the kinds of traits that help creatures survive, reproduce, and generally flourish; this example is directly connected to the story I will relate here. This story concerns my own participation in the science of values over the past decade, endeavors that grew out of a value-centered theory of biological teleology I developed fifteen years ago.

My story starts in 1991 when I was invited to present my theory of teleology to the newly founded artificial life research groups at Los Alamos National Laboratory (LANL) and the Santa Fe Institute (SFI). Artificial life is an interdisciplinary endeavor that studies life and life-like processes by simulating or synthesizing them. Much of this work consists of computer models of processes like the self-organization of simple abstract metabolisms or the evolutionary dynamics of populations of simple self-reproducing automata. Artificial life aims to understand the essential properties of the fundamental processes at work in any possible living system. As Chris Langton once put it, its goal is to understand not "life-as-we-know-it" but "life-as-it-could-be" (Langton 1989). Pursuing this goal requires having a general and broad grasp of what life is and could be, so the LANL-SFI artificial life group created a seminar on the nature of life. But the group was unable to formulate an adequate definition. Disappointed, the best they could produce was a list of characteristic hallmarks of life (see Farmer and Belin 1992).

Teleology in one form or another is often considered one of the hallmarks of life (see, for example, Monod 1971; Mayr 1982), but the notion of teleology is no more self-evident than the notion of life itself. So, knowing about my work, the LANL-SFI artificial life group invited me to present to them a philosopher's perspective on teleology. I knew that the artificial life group did not consider itself to understand a theory fully unless it could see how to implement it in a computer model, so I augmented my presentation with a discussion of how to operationalize the key elements of my theory. My theory of teleology concerns traits that are explained by their value, so operationalizing the theory consisted in figuring out how to determine objectively and impartially when a trait's value or usefulness explains its continued existence. The "objectifying value" of my title refers to this kind of operationalization.

After my lecture, Norman Packard came up and said that he thought it would be easy to objectify teleology in his computer model of sensory-motor evolution. We worked out the details that night and had our first results the next day (see Bedau and Packard 1992). That was an eye-opening episode that convinced me of the usefulness of operationalizing philosophical theories, whenever possible. In this chapter I describe how I objectified value in biology, illustrate the method in a simple evolutionary system consisting of self-replicating computer programs, and explain two fruits of this exercise. One concerns Stephen J. Gould and Richard Lewontin's challenge to adaptationism. The other concerns comparing evolutionary creativity in biological and cultural evolution.

Objectifying Teleological Explanations in Biology

My example of objectifying value in science consists of objectifying the value in a certain kind of biological teleology, specifically, the teleology involved in adaptationist explanations.[1] Traits or behavior that can be explained by reference to the utility of their effects are teleological (*telic*, for the sake of an end), by my lights.[2] In ordinary parlance, telic explanations are offered for a wide variety of things. These include such things as the actions of conscious human agents and the structure and behavior of artifacts designed and used by people. They also include the behavior and structure of biological organisms, as well as certain lower-level components such as genes and also certain higher-level groups such as populations and species. All of these can have telic explanations, and in each case the beneficial effect brought about by the explanandum is an essential part of the explanation. Functionality or adaptiveness is sometimes confused with teleology; the two are related but different. Functional or adaptive behavior is just behavior that is beneficial, that "serves a purpose," regardless of how it comes about. Telic behavior, on the other hand, is not merely beneficial, does not merely serve a purpose. It occurs specifically *because* it is beneficial, *because* it serves a purpose. Telic behavior cannot occur merely accidentally or for some reason unconnected with its utility. Analogous considerations distinguish merely functional or adaptive traits from telic traits.

My concern in the present context is how this framework applies to biological teleology, in particular. A range of behaviors or traits of a given organism at a given time are more or less adaptive. If an organism contains a favorable mutation, the new behavior or trait caused by this mutation might immediately be adaptive or beneficial. But that behavior or trait will not be telic until its utility becomes a causal factor in its continual production. This can happen if its behavior persists through a lineage *because* of its utility.

What I am describing, of course, is the process by which natural selection produces adaptations. An adaptation is a trait (possibly a kind of behavior) that is produced by the process of natural selection *for* that trait.[3] For example, the whale's fins are an adaptation for swimming. The trait persists due to natural selection because of its beneficial effects for swimming; this benefit explains why it is a product of natural selection. Although traits of individual organisms are the paradigm example of adaptations, we can apply the notion to higher level entities by averaging over traits and organisms. In particular, I will talk of genotypes (the complete set of traits in an organism) as adaptations. A genotype is an adaptation if it is persists through the action of natural selection, that is, if on average the individuals with that genotype have been selected for their possession of that genotype, that is, if the traits in that genotype are adaptations.

The crux of my method for objectifying biological teleology is to observe the extent to which items resist selection pressures, for resistance to selection is evidence of adaptation. Since an item is subjected to selection pressure only when it is active or expressed, I call this evolutionary "activity" information.[4] Simple bookkeeping collects a historical record of items' activity — the extent to which items have been subjected to selection pressure, that is, the extent to which their adaptive value has been tested. The bookkeeping incrementally tallies an item's current activity as long as it persists, yielding its *cumulative* activity. If the item (say, a gene) is inherited during reproduction, the offspring's current activity continues to be added to its parent's activity at its birth. In this way our bookkeeping records an item's cumulative activity over its entire history in the lineage. Cumulative activity sums the extent to which an item has been tested by selection over its evolutionary history.

Every time an item is exposed to natural selection, selection can provide feedback about the item's adaptive value. Obviously, an item will not continue to be tested by natural selection unless it has passed previous tests. So, the amount that an item has been tested reflects how *successfully* it has passed the tests. If a sufficiently well-tested item persists and spreads through the population, we have positive evidence that it is persisting *because of* its adaptive value. That is, we have positive evidence that it is an adaptation, that it is telic.

But natural selection is not instantaneous. Repeated trials might be needed to drive out maladaptive items. So exposure to *some* selection is no proof of an item's being an adaptation. Thus nonadaptive items will generate some "noise" in evolutionary activity data. To gauge resistance to selection we must filter out this nonadaptive noise. We can do so if we first measure how activity will accrue to items persisting due just to nonadaptive factors like random drift or architectural necessity. A general way to measure the expected evolutionary activity of nonadaptive items is to construct a *neutral model* of the target system: a system that is similar to the target in all relevant respects *except* that none of the items in it have any adaptive significance. The accumulated activity in neutral models provides a no-adaptation null hypothesis for the target system that can be used to screen off nonadaptive noise. If we observe significantly more evolutionary activity in the target system than in its neutral shadow, we know that this "excess" activity cannot be attributed to nonadaptive factors. It must be the result of natural selection, so the items must be adaptations.[5]

An Illustration of Evolutionary Activity of Genotypes

I will illustrate the evolutionary activity test for adaptations in Evita, a simple artificial evolving system that "lives" in a computer.[6] Somewhat analogous to a population of self-replicating strings of biochemical RNA, Evita consists of a population of self-replicating strings of customized assembly language code that resides in a two-dimensional grid of virtual computer memory. When Evita runs, it doles out CPU time to all programs residing in memory. The system is initialized with a single self-replicating program. CPU time causes this ancestral program to execute, and it copies each of its instructions into a neighboring spot on the grid, thereby producing a new copy of the program—its "offspring." Then this offspring and its parent are both allocated CPU time and start executing, and each makes another copy of itself, creating still more offspring. This process repeats indefinitely. When space in computer memory runs low and offspring cannot find unoccupied neighboring grid locations, the older neighbors are randomly selected and "killed" and the offspring move to the vacated space. Innovations enter the system through point mutations. When a mutation strikes an instruction in a program, the instruction is replaced by another instruction chosen at random, so mutation is continually spawning new kinds of programs.[7] Many are maladaptive, but some reproduce more quickly than their neighbors and these tend to spread through the population, causing the population of strings to evolve over time.

Evita is explicitly designed so that the programs interact only by competing for execution time and space in memory. There is no fixed generation time for Evita programs; some replicate faster than others. On average, those that reproduce faster will supplant their reproducing neighbors. Most significant adaptive events in Evita are changes in reproduction rate, so for present purposes a genotype's fitness can be equated with its reproduction rate. Evita has a clear distinction between genotype and phenotype. A given genotype is simply a string of computer code. If two programs differ in even one instruction they have different genotypes. But two genotypes might produce exactly the same behavior—the same phenotype. If a program includes instructions that never execute, these instructions can mutate freely without affecting the operation of the program. Thus multiple genotypes—without phenotype distinction and so with exactly the same fitness—may then evolve through random genetic drift.

To gather evolutionary activity data in Evita two issues must be settled. First, one must decide which kind of item to observe for adaptations. We will observe whole genotypes. Second, one must operationalize the idea of a genotype's being tested by natural selection. A plausible measure of this is concentration in the population. The greater the genotype's concentration, the more feedback that selection provides about how well adapted it is. A genotype's cumulative evolutionary activity, then, is just the sum of its concentration over time.

In order to discern how much of Evita's genotype activity can be attributed to the genotypes' adaptive significance, we create a "neutral shadow" of it (recall the discussion above). The neutral shadow is a population of nominal "programs" with nominal "genotypes" existing at grid locations, reproducing and dieing. These are not genuine programs with genuine genotypes; they contain no actual instructions. Their only properties are their location on the grid, their time of birth, the sequence of reproduction events (if any) they go through, and their time of death.

Each target Evita run has a corresponding neutral shadow.[8] Certain events in the target cause corresponding events in the shadow, but events in a shadow never affect the target (hence, the 'shadow' terminology). The frequency of mutation events in the shadow is copied from the Evita target. Whenever a mutation strikes a shadow "program" it is assigned a new "genotype." The timing and number of birth and death events in the neutral shadow is also patterned exactly after the target. Shadow children inherit their parent's "genotype" unless there is a mutation, in which case the shadow child is assigned a new "genotype." The key difference is that, while *natural* selection typically affects which *target* program reproduces, *random* selection determines which *shadow* "program" reproduces. So shadow genotypes have no adaptive significance whatsoever; their features like longevity and concentration—and hence their evolutionary activity—cannot be attributed to their adaptive significance. At the same time, by precisely shadowing the births, deaths, and mutations in the target, the neutral shadow shows us the expected evolutionary activity of a genotype in a system exactly like Evita except for being devoid of natural selection. The neutral shadow defines a null hypothesis for the expected evolutionary activity of genotypes affected by only nonadaptive factors such as chance (for example, random genetic drift) or necessity (for example, the system's underlying architecture).

Evita's evolutionary graphs depict the history of the genotypes' activity in a given Evita run. Whenever one genotype drives another to extinction by competitive exclusion, a new wave arises as an earlier one dies out. Multiple waves coexist in the graph when multiple genotypes coexist in the population, and genotypic interactions that affect genotype concentrations are visible as changes in the slopes of waves. The point to appreciate is that the big waves correspond to main adaptations among the genotypes. We can see this clearly in figure 1 by comparing a typical Evita evolutionary activity graph (top) with an activity graph of its neutral shadow (bottom).[9] These graphs are strikingly different.[10] Leaving aside the ancestral wave, the highest waves in the Evita are orders of magnitude higher than those in the neutral analogue. This is clear evidence of how the size of a genotype's evolutionary activity waves in Evita reflects the genotype's adaptive significance. In the Evita target, at each time one or a few genotypes enjoy a special adaptive advantage over their peers, and their correspondingly huge waves reflect this. The change in dominant waves reflects a new adaptation out-competing the prior dominant adaptations. In the neu-

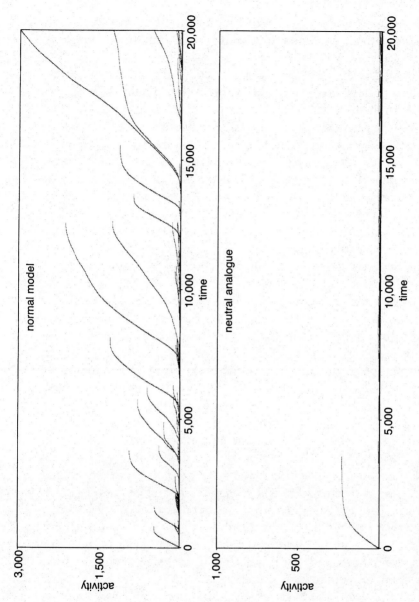

Fig. 1: Evolutionary activity waves in Evita (*top*) and its neutral shadow (*bottom*). Note that the activity scale on the neutral shadow is inflated by a factor of three in order to highlight the neutral shadow waves (*barely visible along the bottom*).

197

tral analogue, by contrast, a genotype's concentration reflects only dumb luck, so no genotype activity waves rise significantly above their peers.

Figure 2 shows more detail of the evolutionary activity during the beginning of the Evita run in figure 1; the average population fitness is graphed in the bottom panel. The activity graph is dominated by five main waves; the first corresponds to the ancestral genotype and the subsequent waves correspond to subsequent adaptations.[11] Miscellaneous low-activity genotypes that never claim a substantial following in the population are barely visible along the bottom of the activity graph. Comparing the origin of the waves with the rises in average population fitness shows that the significant new waves usually correspond to the origin of a higher fitness genotype. Detailed analysis of the specific program that makes up the genotypes with high activity shows that the major adaptive events consist of shortening a genotype's length or copy loop.

The moral, again, is that *significant evolutionary activity waves are significant adaptations*. They correspond to genotypes that are persisting and spreading through the population because of their relative adaptive value. Natural selection is promoting them because of their relative reproduction rate; they flourish because of selection *for* this, so they are adaptations. The evidence for the moral has three parts. First, new significant waves coincide with significant jumps in average population fitness. This shows that the new genotype spreading through the population and making the new wave is an adaptive advantage over its predecessors. Second, microanalysis of the genotypes in the new waves reveals the genetic novelties that create their adaptive advantage. Third, in a neutral model in which chance and architectural necessity are allowed full reign and natural selection is debarred by fiat, no genotypes make significant waves. So, the major evolutionary activity waves in Evita could be produced only by continual natural selection of those genotypes, and natural selection of the genotypes must be due to selection for their adaptive value.

Neutral variant genotypes are an exception to this moral, but they prove the rule. Notice that the second fitness jump in figure 2 corresponds to a dense cloud of activity waves, enlarged in figure 3. The genotypes in this cloud differ from each other only by mutations at an unexpressed locus, so they all use exactly the same algorithm. They are neutral variants of one another — different genotypes with exactly the

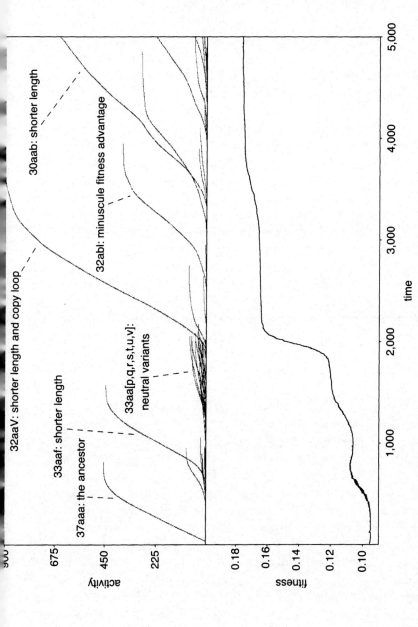

Fig. 2: Evolutionary activity graph (*above*) and average population fitness (*below*) from a typical Evita run. The adaptive advantage of the genotypes causing the salient waves is indicated. The start of a significant wave generally corresponds to an increase in fitness. Note the cloud of neutral variants that causes one of the fitness jumps and acts in the population like a single phenotype. These neutral variants are more fit than genotype 33aaf because they require one fewer instruction per execution of the copy loop. Note also that the significant wave due to genotype 32abl does *not* cause a significant fitness increase; it is nearly phenotypically equivalent to 32aaV because it executes only one fewer instruction per reproduction event than 32aaV.

same phenotype. So the neutral variants are one and the same phenotypic adaptation. Each genotypic instance of the phenotype is an adaptation because it is persisting due to its adaptive value.

The foregoing is an especially simple example of the evolutionary activity test for adaptations, but it is just one example. The test applies to more than just Evita and the like. It can be applied to natural systems as well as artificial systems. For example, it has been applied to evolution in the Phanerozoic age as reflected in the fossil record (Bedau et al. 1997; Bedau, Snyder, and Packard 1998); later in the chapter I will discuss how patent records reflect the evolution of technology. It would be straightforward to apply it to various biological populations, if the relevant raw data could be gathered. For example, Richard Lenski has been directing a study of long-term evolution in *Escherichia coli* (Lenski and Travisano 1994; Cooper and Lenski 2000). Lenski's *E. coli* populations have been evolving for over twenty thousand generations. If we had a record of the concentration of each genotype in these populations every hundred generations or so, we could then plot the evolutionary activity of these genotypes exactly as we did for the genotypes in Evita. Furthermore, it is straightforward to construct a neutral model for the *E. coli* evolution by building a system that copies the key *E. coli* parameters (population size, replication rate, mutation rate, genome size) and driving it with randomly selected births and deaths. So all that stands in the way of measuring evolutionary activity in such natural populations is the practical problem of gathering the raw data.

Evita and *E. coli* are both relatively simple evolutionary systems. Both, for example, are haploid. So one might worry that the evolutionary activity test cannot be applied in more complex contexts. But two considerations defuse this worry. First, those evolutionary systems are not as simple as they might seem at first. Evita, for example, has an indefinitely large space of genetic possibilities, since there is no limit to the length of Evita assembly language programs. And the assembly language is computationally universal as well, so every algorithm whatsoever can be computed by some Evita string. But more importantly, the evolutionary activity test can be naturally extended to apply to many more complicated evolutionary systems. For example, if genetic novelty can be caused by recombination or other more complex genetic operators, this can be handled by including such operators in the neutral model. Or if the environment includes seasonal variation in resources or rare catastrophic changes, these can be added to the neu-

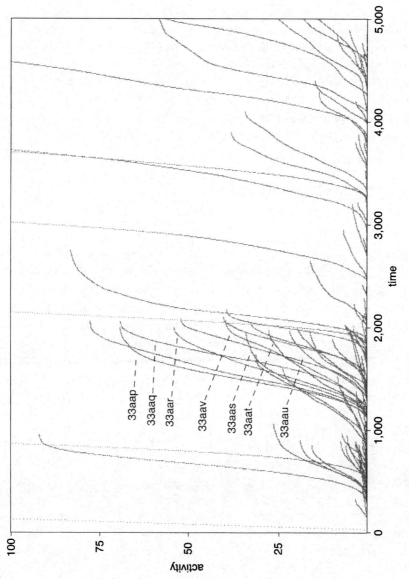

Fig. 3: Blowup of the evolutionary activity graph in fig. 2, showing the neutral variants that cause the fitness increase just after time step 1,000.

tral model. Or if adaptation is reflected in something other than differential reproductive success, then it may well be able to be operationalized in some other way. How exactly to do so would depend on the details of the case at hand. It is impossible to explain how to handle all possible complications in advance, and there is no guarantee that the evolutionary activity test can be extended into all possible contexts. Practical problems might prove insurmountable. My point is that the test easily applies to simple cases and can easily be extended to many more complex ones.

A New Defense of Adaptationism

Adaptive explanations are the bread and butter of evolutionary biology. But the scientific legitimacy of such explanations is controversial, largely because of the classic essay by Gould and Lewontin (1979) "The Spandrals of San Marco and the Panglossian Paradigm: A Critique of the Adaptationist Programme." The controversy persists to this day in large part, I believe, because the fundamental challenge raised by Gould and Lewontin has not yet been met; in fact, it is rarely even acknowledged. The objectification of value in teleology described earlier can change this status quo, for we can now defend the scientific legitimacy of adaptive explanations in a new and deeper way.

First, some terminology. I will refer to claims to the effect that a trait is an adaptation as an *adaptive hypothesis*. A *specific* adaptive hypothesis is a claim to the effect that a trait is an adaptation for some specified adaptive function, and a *general* adaptive hypothesis claims that a trait is an adaptation but identifies no adaptive function. A general adaptive hypothesis expresses the presupposition that the trait has some adaptive explanation. An example is the claim that large primate testes are an adaptation. An example of a specific adaptive hypothesis is the claim that large primate testes are an adaptation for producing more sperm. An *adaptive explanation* of a trait explains its existence or persistence as a result of adaptive evolution, that is, by means of natural selection for that trait. Finally, by *adaptationism* I mean the thesis that the activity of pursuing adaptive explanations of the existence and nature of biological traits is a normal and legitimate part of empirical science.[12]

Gould and Lewontin's central complaint about adaptive explanations is that we have no principled way to tell when they are needed.

People deploy adaptive explanations without justifying them over non-adaptive alternatives, such as appeals to architectural constraints or genetic drift. If one adaptive explanation fails, it is simply replaced by another, but sufficient ingenuity enables any trait to be given an adaptive explanation. The general adaptive hypothesis that a trait *is* an adaptation is treated as untestable. As Lewontin puts it, "the adaptationist program makes of adaptation a metaphysical postulate that ... cannot be refuted" because the presupposition that a trait is an adaptation is never questioned (Lewontin 1977/1985, 76). The deeper worry is that the presupposition that a trait is an adaptation *is* untestable.[13] There is a thicket of alternatives to adaptive explanations. How in principle can we tell when it is appropriate to pursue adaptationist branches? Gould and Lewontin summarize the predicament thus: "We would not object so strenuously to the adaptationist programme if its invocation, in any particular case, could lead in principle to its refutation for want of evidence. We might still view it as restrictive and object to its status as an argument of first choice. But if it could be dismissed after failing some explicit test, then alternatives would get their chance" (Gould and Lewontin 1979, 258ff).

The fundamental challenge, then, is to find some empirical test for general adaptive hypotheses. Without such a test, how can the practice of giving adaptive explanations be a normal and legitimate part of empirical science? In other words, the thesis of adaptationism would seem to be false.

Gould and Lewontin's challenge to adaptationism provoked a storm of response. So many of the responses share the same basic form that this form can be called the "canonical" response. In a nutshell, the canonical response is to concede that there is no general empirical test for general adaptive hypotheses but construe this as on a par with normal empirical science.

Richard Dawkins nicely illustrates the canonical response when he considers traits that might not be adaptations. He points out that it is possible to test rival adaptive hypotheses by ordinary scientific methods, noting that "hypotheses about adaptation have shown themselves in practice, over and over again, to be easily testable, by ordinary, mundane methods of science" (Dawkins 1983b, 360ff). Dawkins's central point is that specific adaptive hypotheses have observable consequences, so they entail empirical predictions and thus can be tested. Dawkins illustrates this point with primate testes size. As it happens, primate testes size scales roughly but not exactly with body size. If testes weight is plotted against body weight, there is considerable scatter around the average line. According to Dawkins,

> A specific adaptive hypothesis is that in those species in which females mate with more than one male, the males need bigger testes than in those species in which mating is monogamous or polyganous: A male whose sperms may be directly competing with the sperms of another male in the body of a female needs lots of sperms to succeed in the competition, and hence big testes. Sure enough, if the points on the testis-weight/body-weight scattergram are examined, it turns out that those above the average line are nearly all from species in which females mate with more than one male; those below the line are all from monogamous or polygynous species. The prediction from the adaptive hypothesis could easily have been falsified. In fact it was borne out. (Dawkins 1983, 361)

This illustrates how specific adaptive hypotheses can be tested by ordinary empirical methods.

Note, though, that Dawkins does not address the testability of *general* adaptive hypotheses. Furthermore, the test for specific adaptive hypotheses cannot be used to produce a test for general adaptive hypotheses. The observable consequences of a specific adaptive hypothesis depend on the specific function hypothesized. Different functions may well entail different predictions. For example, the hypothesis that large primate testes are an adaptation for temperature regulation would entail a quite different prediction about where species fall in the testis-weight/body-weight scattergram. By contrast, the general hypothesis that large testes are an adaptation for something or other entails no prediction about where species fall in the scattergram. So, a *general* adaptive hypothesis inherits no observational consequences from specific hypotheses. For this reason, Dawkins admits that general adaptive hypotheses *are* untestable: "It is true that the one hypothesis that we shall never test is the hypothesis of no adaptive function at all, but only because that is the one hypothesis in this whole area that really *is* untestable" (Dawkins 1983, 361). In other words, Dawkins thinks the fundamental challenge to adaptationism cannot be met. Of course, evidence for a *specific* function is even more conclusive evidence for *some* function, so corroborating a specific adaptive hypothesis also corroborates the corresponding general hypothesis. But we cannot test all possible specific adaptive hypotheses for a trait. So the testability of specific hypotheses provides no test for general hypotheses.

The canonical response is weak.[14] It capitulates in the face of Gould and Lewontin's fundamental challenge by agreeing that there is no test for general adaptive hypotheses. But plenty of traits are not adaptations, and adaptive explanations are often inappropriate. Is there really no empirical way to tell whether adaptive explanations are in the offing? I think the answer is yes and the objectification of biological teleology provides the key. The test we need is simply to collect and analyze evolutionary activity information.

The sign that an evolutionary process is creating adaptations is that its activity data are significantly higher than what would be expected if selection were random. If activity waves rise above the noise generated in a no-adaptation neutral model, then you know the corresponding items are adaptations even if you are ignorant about the adaptive functions. The activity data show that *some* adaptive explanation is needed even if the data are silent about the merits of any *specific* explanation.[15] In other words, the evolutionary activity method tests general rather than specific adaptive hypotheses.

Thus, the evolutionary activity test directly responds to the fundamental challenge to adaptationism. It parts company with the canonical response by not capitulating to Gould and Lewontin. As far as I know, it is the first response that takes this bull by the horns. The test does not assume that traits are adaptations but tests whether they are. Adaptive "just-so" stories have no place here; such stories propose *specific* adaptive hypotheses and these are not at issue. The issue is *general* adaptive hypotheses, and these are accepted only skeptically. Where the canonical response is weak, the activity test is strong. It makes the question of adaptation objective and empirical. When the adaptive stance is adopted, it is on the basis of empirical evidence against nonadaptive alternatives. So we can pursue the adaptationist program constructively and self-critically, as a normal and legitimate part of empirical science. Gould and Lewontin said that they "would not object too strenuously to the adaptationist programme if its invocation, in any particular case, could lead in principle to its refutation for want of evidence" (Gould and Lewontin 1979, 258–59). The evolutionary activity test provides just the sort of tool that Gould and Lewontin sought. So if we can take them at their word, they should now withdraw their objection.

Adaptationism is very controversial, so it is important to defuse some possible misunderstandings about the activity test. For example, some people equate adaptationism with the view that organisms are molded by their environment. However, the activity test does not presuppose this. The test applies perfectly well if the environment is significantly constructed by the organisms in the population. Furthermore, many who defend adaptationism also defend what could be called "panadaptationism"—the view that most (nonmolecular) traits of most organisms are due mostly to natural selection. The activity test takes no stand on panadaptationism. In particular, passing the activity test shows only that natural selection plays some role in the explanation of the trait in question. Other evolutionary forces like drift or architectural necessity might also be important factors in bringing about the trait; the point is that those nonadaptive forces are not sufficient to explain the trait if the evolutionary activity test is positive. Panadaptationists view evolution primarily as a process of adaptive hill climbing. Evolutionary activity measurements could actually be used to argue against this perspective. In effect, I sketch one such argument in the next section, where I suggest that the evolutionary activity signature of the continual production of evolutionary novelties cannot be explained merely by natural selection. That kind of evolutionary hypercreativity depends on some more complicated mechanism than natural selection. We can start to appreciate this by comparing the evolutionary activity in biological and cultural evolution.

Toward a Comparison of Biological and Cultural Evolution

The information in evolutionary activity graphs can be summarized with statistics that reflect how evolution is creating adaptations. Such statistics have various uses, such as enabling quantitative comparisons of adaptive evolution across different systems. After informally explaining some of these statistics and explaining how they have been used to classify evolving systems, I will show how they shed new light on the relationship between biological and cultural evolution.[16]

When attempting to measure the degree of adaptive evolution in a system, one might try to reflect at least three different things: *how well* adapted the adaptations are, that is, how optimally they perform their function; the *intensity* of adaptive evolution, that is, the rate at which new adaptations are being produced by natural selection; and the *extent* of adaptive evolution, that is, the total continual adaptive success of all the adaptations in the system. I will concentrate on the second and third ideas.

The intensity of evolutionary activity intuitively corresponds to the rate that new evolutionary activity is being created in the system, measured as the rate at which new activity waves are entering the activity graph. When there are very few new waves, the intensity of activity is low; when a lot of new waves are being generated, the intensity is high. To clean up this measure of the intensity of evolutionary activity, one would normalize the intensity observed in the target system with the intensity observed in a neutral model, yielding the excess intensity. One simple way to accomplish this is to measure not new waves but new waves that accrue more activity than would be expected in a neutral model.

The *extent* of evolutionary activity intuitively corresponds to the amount of evolutionary activity present in the system, measured as the sum total of the activity in an activity graph at a given time. If you think of the activity waves as being made up of grains of sand, the extent of activity at a given time is the mass of sand at that time in the graph, where the mass is weighted by its height in the graph. When the system has lots of very large activity waves, the extent of activity is very high. When the system has only a few waves and they are relatively low, the extent of activity is relatively low. As with intensity, one would clean up this measurement by normalizing the extent of activity observed in the target system with the extent observed in a neutral model, perhaps simply by subtracting the neutral extent from the target extent, thus yielding the excess extent.

The extent and intensity of evolutionary activity are two independently varying aspects of a system's adaptive evolution. For example, if adaptations continue to persist indefinitely without changing and no new adaptive innovations invade the system, then the extent of activity will continually increase, but the intensity of activity will fall to nil. On the other hand, if evolution is continually creating new adaptations and destroying older ones, the intensity of activity will be positive but the extent of activity will be very low.

The intensity and extent of activity statistics are quite general and apply to data generated by both artificial and natural systems, and they apply at different levels of analysis. I have used evolutionary activity statistics to measure the creation of adaptations in a variety of evolutionary system (Bedau and Packard 1992; Bedau 1995; Bedau 1996; Bedau and Brown 1997; Bedau, Joshi, and Lillie 1999; Bedau et al. 1997; Bedau, Snyder, and Packard 1998; Rechtsteiner and Bedau

1999a,b). Comparing data from a variety of different systems suggests that these statistics can be used to partition evolutionary dynamics into four qualitatively different classes. Class 1 consists of systems in which evolution creates no adaptations at all (for example, all neutral models, systems in which the mutation rate is too high, and systems in which the selection pressure is too low). The signature for this class is zero excess intensity and extent of activity. Systems in which evolution has created adaptations but in which no new adaptations are being created fall into class 2 (for example, stable ecosystems), with the signature of zero excess intensity and unbounded excess extent. Class 3 consists of systems that continually create new adaptations but are bounded in the amount of adaptive structure they contain (for example, if new adaptations always supplant old adaptations). Its signature is positive excess intensity and bounded excess extent. If new adaptations are continually created and the total amount of adaptive structure continues to grow, then the system falls into class 4, which has the signature of positive excess intensity and unbounded excess extent. The biosphere as reflected in the fossil record exhibits class 4 dynamics. (For more details about this classification, see Bedau, Synder, and Packard 1998; and Skusa and Bedau 2002.)

Class 4 is an especially explosive kind of evolutionary creativity. It is intriguing in part because no known existing artificial evolving system generates the same kind of behavior (Bedau et al. 1997; Bedau, Snyder, and Packard 1998). Although we do not know the mechanism behind class 4 behavior, it seems to involve the course of evolution continually creating new kinds of environments that open the door to qualitatively new kinds of adaptations.[17] There is some reason to think that a similar hypercreative process might be at work in cultural evolution. We could start to assess this conjecture if we could apply evolutionary activity statistics to cultural evolution. I have recently started to do this in collaboration with Andre Skusa. Specifically, we have examined the evolution of technology as reflected in patent records, and we use evolutionary activity to create an empirical picture of the adaptive dynamics in patented inventions. Such pictures allow us to compare the dynamics of patented technology with those exhibited in biological evolution.[18]

Patents offer some important advantages for those looking for cultural evolution in empirical data. It is often difficult to operationalize the units of cultural evolution. It is difficult to distinguish new innovations from copies of old innovations when the items are ideas or other mental aspects of culture. Another difficulty is ascertaining precise genealogical relationships. One can finesse these difficulties by studying the evolution of technology as reflected in patent records. Although the evolution of inventions involves the diffusion and selection of ideas, one can identify individual inventions with individual patents. To be patentable an invention must meet three criteria: novelty, usefulness, and nonobviousness. So patented inventions are certified to be new and functional. A patent's novelty is documented by citing the previous patents (and sometimes published papers) that involve related ideas; these are called the patent's "prior art." The citations should identify all the important prior art from which the invention is derived, and in the aggregate they allow a patent's entire genealogy to be inferred.

The analogies and disanalogies between biological and cultural evolution are a matter of some controversy (Hull 1988, 2001), but it is relatively straightforward to extract evolutionary activity data from patent records. The units of evolution we are concerned with (at least in the first instance) are individual patents; these are analogous to genes (or, as memeticists might suggest, "memes"). A gene could vanish forever from an evolutionary system. By contrast, a patented invention never goes fully extinct because the invention exists forever in the patent records. We consider that a patent "reproduces" when it leads to the production of *other* patents; that is, in contrast to most biological evolution, patent reproduction necessarily involves evolutionary innovation.

Especially successful or valuable patents tend to be those that are especially heavily cited. A large body of work in scientometrics has repeatedly confirmed that number of citations is a good reflection of the technological significance and economic value of a patented invention (Albert et al. 1991; Narin 1994; Pavitt 1985; Perko and Narin 1997; Albert 1998). Once a patent has received more than ten citations, the economic value reflected by each additional citation has been estimated to be more than $1 million dollars (Harhoff et al. 1999). For these reasons our bookkeeping of an individual patent's evolutionary activity is based on summing the citations the patent has received. From this perspective, the adaptive success of a patented innovation is measured by the extent to which it spawns subsequent patented innovations. More specifically, we add to a patent's activity at a given time by the number of citations it receives from patents issued at that time.

The fact that a patent has received a few citations does not prove that the invention significantly shapes the evolution of subsequent inventions. A patent might be cited by one or two subsequent patents even if patents to cite were chosen entirely at random. As with evolutionary activity measurements in other contexts, we can evaluate a patent's adaptive significance by comparing its activity with the activity observed in a neutral model of patent evolution. Our patent neutral model mirrors a few key aspects of the real patent data. In both, the same number of patents are issued each week, and they exhibit the same distribution into the various patent classes. Patent citations refer to the same number of pre–September 1996 and post–September 1996 patents, and the references to post–September 1996 patents fall into the various patent classes according to the same distribution. The key distinguishing feature of the neutral model is that the patents to be cited are always chosen randomly.

Skusa and I studied the evolution of technology as reflected in the 868,535 utility patents granted by the United States Patent and Trademark Office between September 1996 and July 2002. Figure 4 shows the dramatic difference between the activity accrued by the most heavily cited real patents and the most heavily cited shadow patents. Overall, the citation levels of shadow patents are very much lower than the citation levels of patents with excess activity. This shows that high citation levels are not due to chance but reflect an invention's value; that is, an invention's salient "reproductive" activity is caused by selection for the invention because of its technological value.

Note that the activity of one patent in figure 4 stands far above the rest, accruing almost twice as many citations as any other patent. This patent covers the technology that allows web browsers to display information, such as advertisements, when a link is clicked while a page is loading. The second most heavily cited patent covers the technology that allows cell phones to receive e-mail and faxes, and the third most heavily cited patent allows remote control of the receipt and delivery of wireless and wireline voice and text messages. All of the ten most heavily cited patents fall into the information technology sector, and seven of them involve the Internet.

More detailed information can be extracted from the evolutionary activity data (see Skusa and Bedau 2002). The point here is simply that evolutionary activity statistics make it feasible to visualize and quantitatively assess the adaptive evolutionary dynamics exhibited in cultural evolution. We have applied the method to technological evolu-

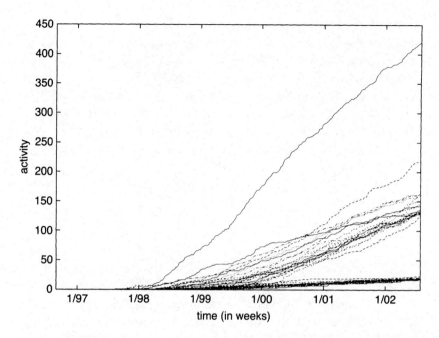

Fig. 4: Comparison of the activity of real patents and shadow patents, showing the twenty most heavily cited patents and the twenty most heavily cited shadow patents. The activity waves of the real patents all rise above a hundred while those of the shadow patents remain below twenty-five. Note that the activity accrued by significant real patents can vastly exceed that accrued by any shadow patent.

tion as reflected in patent record data, but it can be applied to a variety of other cultural systems. Our pilot project underscores the vast importance of information technology, especially the Internet, over the past five years. This is not news, of course; it just corroborates what we already knew. But it does confirm the aptness and probity of evolutionary activity analysis of cultural evolution. Furthermore, it opens the door to quantitative comparison of cultural and biological evolution. And this provides a constructive empirical route for investigating whether the hypercreativity exhibited by biological evolution also characterizes cultural change.

Conclusions

David Hull probably speaks for most philosophers and scientists when he says he wants to "avoid the use of such problematic notions as 'benefit' " in his treatment of natural selection because "their elimination from explanations of biological adaptations was one of Darwin's major achievements" (Hull 2001, 57). I want to counter this attitude with a gestalt switch. In my view, Darwin does not remove value notions like benefit from our understanding of biological adaptation; he simply spells out the objective signs that reflect when those benefits play a certain kind of explanatory role. The gestalt I recommend is to treat Darwin's achievement not as the elimination of value in biology but as its objectification or operationalization.

I have tried to spell out some of the fruits of one case of this sort of objectification of value in biology. That we can objectify values in science is no surprise. We do it all the time. One objectifies the aggregate economic value of a country's commerce by measuring its gross national product, and one objectifies the social values of an individual or social group through appropriate public opinion surveys. Whether such objectifications of values in science are interesting depends mainly on the soundness of the objectification methodology and on the specific insights revealed (if any). The proof of this pudding is in the eating. If my case of objectification has any significant interest, it lies in the evolutionary activity statistics and their application in the new defense of adaptationism and the new ability to compare biological and cultural evolution. The main moral I draw from this one case study is that my value-centered view of teleology would never have born these fruits but for its objectification.

NOTES

Special thanks to my collaborators on the evolutionary activity investigations reported here: Titus Brown, Norman Packard, and Andre Skusa. Special thanks also to James Lennox for his helpful prepared commentary on a version of this chapter presented at the Sixth Pittsburgh-Konstanz Conference in the Philosophy of Science. Thanks for helpful discussion to the audience in that conference, especially David Griffiths, Peter Machamer, Sandra Mitchell, and Jerome Shaffner. For helpful comments or discussion on earlier versions of parts of this chapter, thanks to Phil Anderson, Peter Godfrey-Smith, Mike Raven, Tom Ryckman, and Chris Stephens. Thanks also for helpful discussion to audiences at the University of Oklahoma, the University of Washington, Washington University in St. Louis, the Center for Humanities at Oregon State University, the Center for Cognitive Studies at Tufts University, the Santa Fe Institute, the first Genetic and Evolutionary Computation Conference, the fourth European Conference on Artificial Life, the sixth Artificial Life Conference, the Lake Arrowhead Conference on Computational

Social Science, and the Fraunhofer Gesellshaft in Sankt Augustin, Germany, where some of the material here was presented. And thanks to the University of Oklahoma, its Zoology Department, and Professor Tom Ray, for hospitality and support while some of this work was accomplished.

1. Another interesting form of biological teleology has to do with the purported inherent "progress" exhibited by the course of evolution. For a discussion of how my objectification of value in biology bears on this, see Bedau (1998).

2. For more on my value-centered view of teleology, and for comparisons with other views of teleology, see Bedau (1990, 1991, 1992a,b, 1993).

3. Not just selection *of* the trait; for this distinction see Sober (1984).

4. "Selection test" information would be more informative (and more awkward) terminology. Norman Packard and I developed and applied this method to a number of systems over a number of years with the help of students and colleagues. See Bedau and Packard (1992), Bedau (1995, 1996), Bedau and Brown (1997), Bedau et al. (1997), Bedau, Snyder, and Packard (1998), Bedau, Joshi, and Lilly (1999), Rechtsteiner and Bedau (1999a,b).

5. Although the evolutionary activity method is novel, the essential logic behind it should be familiar. See, for example, Kimura (1983), Wimsatt (1987), Raup (1987), Beatty (1987). These parallels are traced in greater detail in Bedau (2002).

6. Created by C. Titus Brown, Evita is inspired by Tierra (Ray 1992) and its derivative Avida (Adami and Brown 1994), but it is much simpler because it disallows the kind of interactions that lead to parasitism and the other interesting evolutionary phenomena observed in Tierra. Its simplicity makes it an especially simple and clear illustration of how graphing evolutionary activity reveals a system's evolutionary dynamics. A much more detailed presentation of the material in this section is available elsewhere (Bedau and Brown 1997).

7. If the mutation rate is too low, there is no significant genetic change in the population. If the mutation rate is too high, the population dies out almost immediately because no successfully reproducing creature can survive the bombardment of mutations long enough to reproduce.

8. Actually, it has an indefinite number of them, due to random sampling differences — a qualification I will usually ignore.

9. In the Evita activity graphs shown here, the Evita system parameters were all identical except for mutation rate and elapsed time. Each genotype in a given run is given a unique name of the form Nxxx, where N is a number indicating the genotype's length and xxx is a three-character string (in effect, a base 52 number) indicating the genotype's order of origination among genotypes of that length. For example, 32aac is the third length 32 genotype to arise in the course of a given run.

The grid size was forty by forty, so when the grid filled up, the population consisted of about sixteen hundred self-reproducing programs. I have pruned out irrelevant data about transitory genotypes by graphing only those genotypes that had at least five instances in the population at some time. This removes some of the "little hairs" created by nonadaptive noise (see fig. 3).

10. Note that the activity scales (vertical axes) in these two plots are roughly comparable, except that in the bottom panel activity is expanded by a factor of three to make the neutral model activity easier to see.

11. Notice that the fourth salient wave (due to genotype 32abl) does *not* corre-

spond to a significant fitness jump. This genotype is well adapted, but it is not significantly better adapted than its main predecessor, genotype 32aaV. The waves from 32aaV and 32abl coexist for so long because the two genotpyes are nearly neutral variants. In fact, the fitness of the second wave (32abl) exceeds that of the first wave (32aaV) by about only 0.5 percent. The interactions among the three salient waves between updates 4,000 and 5,000 have a similar explanation. They are a significant improvement (5 percent fitness advantage) over the genotypes that they drive extinct, but they differ from one another by much less (less than 2 percent).

12. My use of the term *adaptationism* captures what I believe is the central issue, but I should emphasize that the term is sometimes used in other ways. To get a sense of the similarities and differences, see Maynard Smith (1978), Dawkins (1982), Dupré (1987), Sober (1987), Brandon (1990), Burien (1992), West-Eberhard (1992), Orzack and Sober (1994, 2001), Godfrey-Smith (2001).

13. It might be uncontroversial that some specific traits are adaptations, but those are the exception.

14. One can begin to appreciate just how canonical this response is by examining some other well-known responses to Gould and Lewontin (for example, Mayr 1983; Dennett 1983; Rosenberg 1985; Sober 1987; Dennett 1995; Sober 1993). I spell out these parallels elsewhere (Bedau 2002).

15. For example, figure 2 shows that the big wave in the middle produced by genotype 32aaV is an adaptation but it does not show what makes it better than its peers. We can usually discover a genotype's adaptive advantage by independent microanalysis, as we did for 32aaV.

Genetic hitchhikers and genetic drift introduce some complications in this analysis. I discuss these details in a forthcoming monograph.

16. Mathematical details about the statistics can be found elsewhere (Bedau and Packard 1992; Bedau 1995; Bedau et al. 1997; Bedau, Snyder, and Packard 1998).

17. The term *evolutionary novelty* is sometimes used to refer to the production of qualitatively new kinds of adaptations. So the evolutionary hypercreativity of class 4 dynamics involves the continual production of new kinds of evolutionary novelties. Developmental systems theorists sometimes call attention to individual examples of evolutionary novelties. My focus here is complementary but broader: the general pattern of the continual production of new kinds of evolutionary novelties.

18. The material in this section is explained in greater detail in Skusa and Bedau 2002; there is plenty of previous work on cultural evolution and on patents, but none quite like ours. For many years cultural change has been treated as a process of the diffusion of ideas (Rogers 1995), and the scientometrics community has been investigating scientific and technological change by analysis of bibliometric data and patent records for decades (Pavitt 1985; Garfield and Welljams-Dorof 1992; Narin 1994; Albert 1998). But these approaches view "evolution" simply as any change in time rather than just change resulting from differential imperfect replication and selection. Sociobiology (Wilson 1978; Lumsden and Wilson 1981) and its contemporary sibling, evolutionary psychology (Barkow, Cosmides, and

Tooby 1992), explore one kind of connection between biological and cultural evolution, specifically, the extent to which certain psychological and cultural phenomena (such as homosexuality and altruism) can be explained by appealing to the operation of biological evolution itself. This reduction of social science to biology is contrasted with the approach to culture illustrated by memetics (Lynch 1996; Blackmore 1999; Aunger 2000), which considers the evolution of cultural phenomena in its own right, independent from and even competing with biological evolution. The two classic quantitative treatments of cultural evolution (Cavalli-Sforza and Feldman 1981; Boyd and Richerson 1985) tend toward different answers to the question whether cultural evolution is ultimately explainable in terms of biological evolution, with Cavalli-Sforza and Feldman leaning toward explanatory dependence and Boyd and Richerson leaning toward a limited autonomy for culture. My approach is neutral on this issue. I study cultural evolution as an evolutionary process in its own right, ignoring whether and how it might depend on biological evolution. My goal is to provide an empirical and quantitative picture of the evolution of culture, one that allows us to compare its evolutionary dynamics with those of biological evolution. Both reductionists and antireductionists could profit from objective empirical measurement of cultural dynamics.

REFERENCES

Adami, C., and C. T. Brown. 1994. Evolutionary learning in the 2D artificial life system "Avida." In *Artificial life IV*, ed. R. Brooks and P. Maes, 377–81. Cambridge, MA: MIT Press.

Albert, M. B. 1998. *The new innovators: Global patenting trends in five sectors*. Washington, DC: U.S. Department of Commerce.

Albert, M., D. Avery, F. Narin, and P. McAllister. 1991. Direct validation of citation counts as indicators of industrially important patents. *Research Policy* 20:251–59.

Aunger, R., ed. 2000. *Darwinizing culture: The status of memetics as a science*. New York: Oxford University Press.

Barkow, J. H., L. Cosmides, and J. Tooby, eds. 1992. *The adapted mind: Evolutionary psychology and the generation of culture*. New York: Oxford University Press.

Beatty, John. 1987. Natural selection and the null hypothesis. In *The latest on the best: Essays on evolution and optimization*, ed. J. Dupré, 53–75. Cambridge, MA: MIT Press.

Bedau, M. A. 1990. Against mentalism in teleology. *American Philosophical Quarterly* 27:61–70.

———. 1991. Can biological teleology be naturalized? *Journal of Philosophy* 88:647–55.

———. 1992a. Where's the good in teleology? *Philosophy and Phenomenological Research* 52:781–806.

———. 1992b. Goal-directed systems and the good. *Monist* 75:34–49.

———. 1993. Naturalism and Teleology. In *Naturalism: A critical appraisal*, ed. S. Wagner and R. Warner, 23–51. Notre Dame, IN: University of Notre Dame Press.

———. 1995. Three illustrations of artificial life's working hypothesis. In *Evolution and biocomputation: Computational models of evolution*, ed. W. Banzhaf and F. Eeckman, 53–68. Berlin: Springer. Available at http://www.reed.edu/@mab/biocomputation.pdf.

———. 1996. The nature of life. In *The philosophy of artificial life*, ed. M. Boden, 332–57. New York: Oxford University Press. Available at http://www.reed.edu/@mab/life.OXFORD.html

———. 1998. Philosophical content and method of artificial life. In *The digital phoenix: How computers are changing philosophy*, ed. T. W. Bynam and J. H. Moor, 135–52. Oxford: Basil Blackwell.

———. 2002. A new defense of adaptationism. Paper presented at the Northwest Philosophy Conference, Lewis and Clark College, Portland, OR.

Bedau, M. A., and C. T. Brown. 1997. Visualizing evolutionary activity of genotypes. *Artificial Life* 5:17–35. Available at http://www.reed.edu/@mab/vis_gtypes_alifejournal.pdf.

Bedau, M. A., and N. Packard. 1992. Measurement of evolutionary activity, teleology, and life. In *Artificial life II*, ed. C. Langton, C. Taylor, J. D. Farmer, and S. Rasmussen, 431–61. Redwood City, CA: Addison Wesley. Available at http://www.reed.edu/@mab/alife2.pdf.

Bedau, M. A., S. Joshi, and B. Lillie. 1999. Visualizing waves of evolutionary activity of alleles. In *Proceedings of 1999 genetic and evolutionary computation conference workshop program*, ed. A. Wu, 96–98. Orlando, FL: GECCO Proceedings. Available at http://www.reed.edu/@mab/vis_gecco99.pdf.

Bedau, M. A., E. Snyder, and N. H. Packard. 1998. A classification of long-term evolutionary dynamics. In *Artificial life VI*, ed. C. Adami, R. Belew, H. Kitano, and C. Taylor, eds., 228–37. Cambridge, MA: MIT Press. Available at http://www.reed.edu/@mab/alife6.pdf.

Bedau, M. A., E. Snyder, C. T. Brown, and N. H. Packard. 1997. A comparison of evolutionary activity in artificial systems and in the biosphere. In *Proceedings of the fourth European conference on artificial life, ECAL97*, ed. P. Husbands, and I. Harvey, 125–34. Cambridge, MA: MIT Press. Available at http://www.reed.edu/@mab/ecal97.pdf.

Blackmore, S. 1999. *The meme machine*. New York: Oxford University Press.

Brandon, R. N. 1990. *Adaptation and environment*. Princeton, NJ: Princeton University Press.

Boyd, R., and P. J. Richerson, 1985. *Culture and the evolutionary process*. Chicago, IL: University of Chicago Press.

Burien, Richard M. 1992. Adaptation: Historical perspectives. In *Keywords in evolutionary biology*, ed. E. F. Keller and E. A. Lloyd, 7–12. Cambridge, MA: Harvard University Press.

Cavalli-Sforza, L. L., and M. W. Feldman. 1981. *Cultural transmission and evolution: A quantitative approach*. Princeton, NJ: Princeton University Press.

Cooper, V. S., and R. E. Lenski. 2000. The population genetics of ecological specialization in evolving *Escherichia coli* populations. *Nature* 407:736–39.

Dawkins, R. D. 1982. *The extended phenotype: The long reach of the gene.* New York: Oxford University Press.
———. 1983. Adaptationism was always predictive and needed no defense. Commentary on Dennett. *Behavioral and Brain Sciences* 6:360–61.
Dennett, D. C. 1983. Intentional systems in cognitive ethology: The "panglossian paradigm" defended. *Behavioral and Brain Sciences* 6:343–55.
Dennett, D. 1995. *Darwin's dangerous idea: Evolution and the meanings of life.* New York: Simon and Schuster.
Dupré, J., ed. 1987. *The latest on the best: Essays on evolution and optimization.* Cambridge, MA: MIT Press.
Farmer, D., and A. Belin. 1992. Artificial life: The coming evolution. In *Artificial life II*, ed. C. Langton, C. Taylor, J. D. Farmer, and S. Rasmussen, 815–40. Redwood City, CA: Addison Wesley.
Garfield, E., and A. Welljams-Dorof. 1992. Their use as quantitative indicators for science and technology evaluation and policy making. *Science and Public Policy* 19:321–27.
Godfrey-Smith, P. 2001. Three kinds of adaptationism. In *Adaptationism and optimality*, ed. S. H. Orzack and E. Sober, 335–57. Cambridge: Cambridge University Press.
Gould, S. J., and R. C. Lewontin. 1979. The spandrels of San Marco and the panglossian paradigm: A critique of the adaptationist programme. *Proceedings of the Royal Society* B205:581–98. Reprinted in *Conceptual issues in evolutionary biology*, ed. E. Sober, 73–90. Cambridge, MA: MIT Press, 1993. Page references are to the 1979 edition.
Harhoff, D., F. Narin, F. M. Scherer, and K. Vopel. 1999. Citation frequency and the value of patented innovation. *Research Policy* 81:511–15.
Hull, D. L. 1988. *Science as a process: An evolutionary account of the social and conceptual development of science.* Chicago, IL: University of Chicago Press.
———. 2001. *Science and selection: Essays on biological evolution and the philosophy of science.* Cambridge: Cambridge University Press.
Kimura, M. 1983. *The neutral theory of molecular evolution.* Cambridge: Cambridge University Press.
Langton, C. G. 1989. Artificial Life. In *Artificial Life*, ed. C. G. Langton, 1–47. Redwood City, CA: Addison-Wesley.
Lenski, R. E., and M. Travisano. 1994. Dynamics of adaptation and diversification: A 10,000-generation experiment with bacterial populations. *Proceedings of the National Acadamy of Sciences, USA* 91, 6806–14.
Lewontin, Richard. 1977/1985. Adaptation. In *The dialectical biologist*, ed. R. Levins and R. Lewontin. Cambridge, MA: Harvard University Press. Originally published as "Adattamento" in *Enciclopedia Einaudi*, vol. 1, ed. G. Einaudi. Turin, Italy, 1977.
Lumsden, C. J., and E. O. Wilson. 1981. *Genes, mind, and culture.* Cambridge, MA: Harvard University Press.
Lynch, A. 1996. *Thought contagion: How belief spreads through society.* New York: Basic Books.
Maynard Smith, J. 1978. Optimization theory in evolution. *Annual Review of Ecology and Systematics* 9:31–56.

Mayr, E. 1982. *The growth of biological thought*. Cambridge, MA: Harvard University Press.
———. 1983. How to carry out the adaptationist program? *American Naturalist* 121 (March):324–33. Reprinted in *Toward a new philosophy of biology: Observations of an evolutionist*, ed. E. Mayr, 148–59. Cambridge, MA: Harvard University Press, 1988. Page references are to the 1988 edition.
Monod, J. 1971. *Chance and necessity*. New York: Vintage.
Narin, F. 1994. Patent bibliometrics. *Scientometrics* 30:147–55.
Orzack, S. H., and E. Sober, eds. 2001. *Adaptationism and optimality*. Cambridge: Cambridge University Press.
Orzack, S. H., and E. Sober. 1994. Optimality models and the test of adaptationism. *American Naturalist* 143:361–80.
Pavitt, K. 1985. Patent statistics as indicators of innovative activities: Possibilities and problems. *Scientometrics* 7:77–99.
Perko, J. S., and Narin, F. 1997. The transfer of public science to patented technology: A case study in agricultural science. *Journal of Technology Transfer* 22: 65–72.
Raup, D. M. 1987. Neutral models in paleobiology. In *Neutral models in biology*, ed. M. H. Nitecki and A. Hoffman, 121–32. New York: Oxford University Press.
Ray, T. S. 1992. An approach to the synthesis of life. In *Artificial life II*, ed. C. Langton, C. Taylor, J. D. Farmer, S. and Rasmussen, 371–408. Redwood City, CA: Addison Wesley.
Rechtsteiner, A., and M. A. Bedau. 1999a. A generic model for quantitative comparison of genotypic evolutionary activity. In *Advances in artificial life*, ed. D. Floreano, J.-D. Nicoud, F. Mondada, 109–18. Heidelberg: Springer-Verlag. Available at http://www.reed.edu/@mab/ecal99.pdf.
Rechtsteiner, A., and M. A. Bedau. 1999b. A generic model for measuring excess evolutionary activity. In *GECCO-99: Proceedings of the genetic and evolutionary computation conference*, vol. 2, ed. W. Banzhaf, J. Daida, A. E. Eiben, M. H. Garzon, V. Honavar, M. Jakiela, and R. E. Smith, 1366–73. San Francisco, CA: Morgan Kaufmann. Available at http://www.reed.edu/@mab/gecco99.pdf.
Rogers, E. M. 1995. *Diffusion of innovations*. 4th ed. New York: Free Press.
Rosenberg, A. 1985. *The structure of biological science*. Cambridge: Cambridge University Press.
Skusa, A., and M. A. Bedau. 2002. Toward a comparison of evolutionary creativity in biological and cultural evolution. In *Artificial life VIII*, ed. R. Standish, M. A. Bedau, and H. Abbass, 233–42. Cambridge, MA: MIT Press.
Sober, E. 1984. *The nature of selection: Evolutionary theory in philosophical focus*. Cambridge, MA: MIT Press.
———. 1987. What is adaptationism? In *The latest on the best: Essays on evolution and optimization*, ed. J. Dupré, 105–118. Cambridge, MA: MIT Press.
———. 1993. *Philosophy of biology*. Boulder, CO: Westview Press.
West-Eberhard, M. J. 1992. Adaptation: Current usages. In *Keywords in evolutionary biology*, ed. E. F. Keller and E. A. Lloyd, 13–18. Cambridge, MA: Harvard University Press.

Wilson, E. O. 1978. *On human nature*. Cambridge, MA: Harvard University Press.

Wimsatt, W. C. 1987. False models as means to truer theories. In *Neutral models in biology*, ed. M. H. Nitecki and A. Hoffman, 23–55. New York: Oxford University Press.

11

Border Skirmishes between Science and Policy:

Autonomy, Responsibility, and Values

Heather E. Douglas
Department of Philosophy, University of Tennessee

Introduction

Unlike the academically centered science wars of the 1990s, which appear to be petering out, the debate over the role of science in policymaking continues unabated since (at least) the 1970s. While many philosophers of science look to the creationism debates and science test scores to bemoan the decline of science in American society, policymakers' reliance on science has never been greater. Thousands of scientists participate in the shaping of policy at the federal level through both legislatively mandated and less formal committees. Science has become such an entrenched part of public policymaking that there is little chance science will be rejected as an authoritative voice in that venue. While the necessity of science in policymaking is accepted, what is regularly debated is what the precise role of science should be in the policymaking process.

A central part of this debate over science's role is the continually renewed attempt to erect a clear boundary between science and policy/politics. Such a boundary is viewed as good for both science and policy. Science benefits by keeping potentially corrupting political influences (and their attached social values) from the practice of science, thus keeping science "pure" (or at least autonomous). Policy benefits by keeping the political responsibility for the final decision in the hands of elected officials (or at least public officials accountable to the electorate), and out of the relatively small group of experts from whom they receive advice. In practice, however, defining and maintaining a clear boundary between these two areas has proven notoriously difficult.

In this chapter, I will review the trajectory of these skirmishes over where science stops and policy begins. The battles present us with important lessons. I will argue that a clear boundary between science and policy is not only difficult to maintain in practice but also normatively undesirable if the boundary is defined in terms of the role of values, as has historically been the case. Because of the necessity of publicly responsible value judgments in the practice of policy-relevant science, no clear demarcation can or should be made between the two in terms of the role of values. Attempting to avoid this necessity, I will claim, is a greater threat to science's autonomy than fully accepting and acknowledging both it and its attendant responsibility.

Two caveats before I begin. First, I will be focusing here on the use of science in policymaking, not on the development of policy for science. While all who have used this classic divide acknowledge its artificiality (because of the interplay between the two areas), it is a useful way to narrow the scope of this project to a manageable size. Second, I will not attempt to provide a complete history of U.S. science advice here. To my knowledge no such history has yet been written, although it would be very useful. Thus, the periodization I present here is a sketch and will likely need future modification. What I hope to show with the sketch is how the issue of values in science has been central to the attempts at boundary definition. As we will see, the failures of these definitions are made understandable once we revise our views on the role of values in science.

Boundaries: A History of Science Advice, 1945–2000

Examples of successful science advice for policy prior to World War II are sparse. Science was utilized by the government in isolated areas (agriculture, exploration, standards) and during war times (witness the formation of the National Research Council to expand the National Academy of Science's capabilities during World War I), but attempts to gain scientific input for policymaking in general, on a regular basis, or at higher policy levels were not successful. For example, the National Academy of Sciences, formed during the Civil War to advise the government, became an honorary society after the end of the war. At the end of World War I, the newly formed National Research Council drifted away from its advisory role. (For details on this period, see Dupre 1957.) The Depression Era Science Advisory Board was largely viewed as a failure, even by contemporaries (Pursell 1965). World War II, however, was to change all this. Vannevar Bush's brilliant construction of a framework within which science and government could work closely together and still tolerate each other changed the traditionally frosty relationships between the scientific community and government officials. At the end of the war, there was no debate over whether some close relationship (including federal funding) was needed between science and government. The only debates were over the form of that relationship. It is for this reason that World War II is seen as a watershed event in the history of American science policy.

Two key developments during World War II influenced all the science policy that was to follow. The first was the recognition by government officials of how tremendously useful scientists could be. The impact of science on the trajectory of the war was far greater than in any earlier war. While the atomic bomb might get the most attention from our current perspective, developments from penicillin to short wave radar all did much to change the way in which we waged war, and may well have changed the outcome of the war. Bureaucratic and elected officials agreed for the first time on the tremendous value of science for both the government and the nation.

The second was the development of mechanisms by which scientists could work for the government and still feel they had sufficient autonomy to be functioning scientists. The government contract, administered through an institution to an individual (or set of individuals in a lab), allowed the government to have some ability to hold scientists accountable for their work, while allowing scientists a relatively free hand to pursue the work they found most interesting or important. Contrast this with two other possible avenues of government support for science: the government lab (where accountability could be greater but academic freedom would be less) and the direct grant to universities (where little accountability would be possible). To many the contract grant seemed to strike the right balance between government support and government interference, helping scientists to maintain their autonomy while utilizing an incredible increase in public funding.[1]

What was not developed in World War II was an effective long-term mechanism by which the government (the president, the federal agencies, or Congress) could receive science advice. As with the issue of funding, World War II broke new ground in science-government relationships in the arena of advice. According to Edward Burger, "Vannevar Bush, perhaps, deserves the title of the first real presidential science advisor. . . . The appointment of Bush . . . as the head of the Office of Scientific Research and Development, was the first serious attempt to couple scientific talent to the public policy machinery" (Burger 1980, 6).

While Bush's role with the Office of Scientific Research and Development (OSRD) was a crucial step, it could not serve as a general model. Bush's OSRD was effective because of Bush's good relationship with Franklin D. Roosevelt (and FDR's desire for science advice), because of Bush's foresight to set up the institutional arrangements before the United States entered World War II, and because of the extremities of wartime, which allowed Bush to control large amounts of funding in almost total secrecy. What made the OSRD a successful source of advice during the war could not be replicated after the war.

Yet because of the growing importance of science-based questions in national policy, particularly on issues of defense, and because of the complexity of these policy issues, some source of science advice was needed. The history of science advice after World War II is marked by continual experimentation with various avenues by which the government could glean science advice. Endemic to this history are attempts to define the role of science in policymaking, and in particular to find some way to clearly distinguish science from policy.

The development of science advice in the United States falls into three rough periods: 1945–1965, during which there was an ontologically based distinction between science and policy based on the general belief that scientists' advice was naturally value-free; 1965–1980, a time of theoretical upheaval generated when the illusion of value-free advice fell apart and was replaced by dueling scientists with clearly differing values; and 1980–2000, during which a focus on *processes* developed with the aim of keeping values out of the science to be used in policymaking. Each of these periods has its contemporaneous critics, but I hope to capture the spirit of the time periods rather than a universally held perspective. Throughout these periods, the advisory system has continually expanded and become more institutionally entrenched. We are now faced with an extensive system, much of it mandated by law and governed by both federal law (such as the Federal Advisory Committee Act) and extensive regulatory efforts.

Ontologically Defined Boundary, 1945–1965

The twenty years following World War II are often looked on as the golden age of science-government relations. Government budgets for science continually expanded at an extraordinary pace. Following the vision outlined in Vannevar Bush's *The Endless Frontier*, all basic research was assumed to be valuable to society in general. According to Bush, the benefits science would provide included "more jobs, higher wages, shorter hours, more abundant crops, more leisure for recreation, for study, for learning how to live without the deadening drudgery which has been the burden of the common man for ages past" (Bush 1945/1960, 10) This vision was widely shared and thus government funding of science increased dramatically in the years after the war. Because of wrangling over the shape of a central science funding agency (not whether such funding should occur), the National Science Foundation was not founded until 1950. In the interim years, multiple government sectors filled in the funding gaps. By the mid-1950s, science research funding could be had from the government through the Department of Defense (particularly the Office of Naval Research), the Atomic Energy Commission (AEC), the National Institutes of Health, the Department of Agriculture, the Department of Commerce, the Department of the Interior, the National Advisory Committee on Aeronautics, and the National Science Foundation.[2]

The power of scientists in government rapidly expanded as well, as the government experimented with different institutional arrangements for science advice. Several early advisory committees were set up largely to advise the government on how best to spend money on scientific research. For example, the Joint Research and Development Board was created in 1946 (and then recreated as simply the Research and Development Board in 1947), accompanied by numerous advisory boards, on which scientists sat to help direct research funds for the Army and Navy (Kleinman 1995, 149). However, several bodies appeared that would do more than this, that would shape other areas of policy as well. The earliest example is probably the AEC's General Advisory Committee (GAC). Created through the Atomic Energy Act of 1946 expressly to provide advice to the newly formed AEC, the GAC was a well-respected and influential body of scientists, including the likes of Glenn Seaborg, Lee DuBridge, and Enrico Fermi (Kevles 1971/1995, 377; Hewlett and Duncan 1962/1990, 665). Chaired by

Robert Oppenheimer from 1946 to 1952 and then by I. I. Rabi from 1952 to 1956, the GAC had a strong influence on AEC policy in its first decade. As described by Hewlett and Duncan, even at its very inception, the committee "spoke with the voice of authority. Its distinguished membership would have assured effectiveness in almost any situation; in the absence of a strong Commission leadership in March 1947, the committee's opinions were almost overriding" (Hewlett and Duncan 1962/1990, 46).

Higher level advisory bodies also developed during this period. In 1951, at the advice of William Golding, President Harry S. Truman created the first official presidential science advisory body housed in the executive office, the Science Advisory Committee (Smith 1992, 162–63). Attached to the Office of Defense Mobilization (ODM) and consisting of eleven prominent scientists, it was little utilized by Truman (Smith 1992, 162–63; Burger 1980, 7–8). The ODM committee was more effective under Eisenhower's administration, however, providing important advice on military technology questions such as missile defense (Smith 1992, 164). In 1957, in response to the launch of *Sputnik,* Eisenhower expanded the role of the committee, renaming it the Presidential Science Advisory Committee (PSAC), and appointing the chair of that committee as his Special Assistant on Science (Burger 1980, 7).[3] By 1960, roughly one hundred scientists served in an advisory capacity to the government through PSAC panels (Gilpin 1964, 8, n16).

As the institutional structures for science advice proliferated in the 1950s, the general number of scientists serving in advisory capacities and the time commitment of those scientists also increased greatly. In addition to the formal commitment to sit on a standing advisory body, many scientists participated in the less formal summer studies programs, in which scientists would gather to discuss a particular problem and to provide advice on just that problem (Greenberg 1967/1999, 28). Rather than complaining about being ignored, as they had prior to World War II, some scientists complained about the time advising the government took away from their research. For example, in 1962 Hans Bethe would complain about the amount of time advising the government took away from research, for both the senior statesmen of science like himself and for younger scientists (Wood 1964, 41–42).

Despite this heavy involvement with the government, there was little concern in this period over policy corrupting science/scientists or scientists unduly influencing policy. The general picture was as follows: Because scientists were providing scientific expertise, and because science was free from social and political values, the advice the scientists provided was also free of those values. This "value neutrality" was thought to be a characteristic *inherent* to most scientists and central to the continuation of sound science advice. For example, in describing the selection process for PSAC, Robert Gilpin presents the goal of the administration in this way: "to obtain a highly competent, experienced, and politically neutral body of scientific experts" (Gilpin 1964, 10). The process by which one obtained these neutral advisors was to recruit members on the basis of past successful service on lower level advisory bodies. Because scientists helped select the lower level appointees, and scientists were presumably politically neutral, drawing from these lower level appointees would perpetuate the political neutrality of the science advising process. Thus would Robert Wood label the science advisors an "apolitical elite" (Wood 1964).

More savvy observers of science advising also had difficulty in seeing scientists as pursuing just another set of special interests. In his essay "The Scientific Advisor," Harvey Brooks acknowledges the role that values must play in the giving of science advice: "All recommendations involve nontechnical assumptions or judgments in varying degrees" (Brooks 1964, 76). Nevertheless, there is something special about science advice: "While scientific advice is not free of bias, or even of special pleading, it is probably more free of prejudice than much other professional advice" (80). A belief in the special ability of scientists to be politically neutral was widely held.

Reflection on this period provides at least some explanation for the predominance of this view. Science advisors of this period almost exclusively gave advice on military and national security issues. The most prominent bodies, the GAC and PSAC, were originally created to advise governmental bodies in this area, and during the Cold War period, divergent values on topics related to national security were hard to come by. The political values the science advisors of this period held may have been largely invisible simply because of the broad agreement in society on social and political values. As one contemporaneous commentator noted, the Cold War climate created a "gross simplification of value alternatives" and led to a "decline of ideological conflict" (Wood 1964, 54). If expert judgments are made on the basis of widely accepted values, it will not appear that any values had a role in those judgments at all; instead it will appear simply as though the correct judgment was made. As science advisors moved into areas related to domestic policy in the 1960s, with all its attendant social conflict, the value judgments inherent in advising became more apparent.

By the early 1960s, scientists had made themselves part of the policy process, but had at the same time kept themselves separate from the political fray in the minds of the politicians and the public. Scientists were seen as simply *being* apolitical, of being "neutral" and value-free, and thus the boundary between science and policy was kept safe and secure by the nature of that being. Over the next decade, that view would fall apart. Change was already in the wind when Gilpin wrote, "All agree that that phase of contemporary scientific-government relations which began in 1945 is ending, and that the United States is beginning to experience new problems and challenges in this area of national life" (Gilpin 1964, 3). It was clear that the era of unscrutinized budget expansions for science funding was over and that this might create problems among advisory committees used to evaluate science funding (Brooks 1964, 96). However, the problems would go much deeper than conflict among scientists over dollars.

Dissolution of Boundaries, Expansion of Institutions, 1965–1980

By the mid-1960s, cracks began to form in this benign view of science involvement with policy. The issues on which science advisors gave advice expanded greatly, moving beyond the traditional realm of military technology and national security to include public health, transportation, consumer safety, and environmental issues. At the same time, public opinions and values on policy issues sharply diverged. The Vietnam War and civil rights struggle set a backdrop of general societal contention over major issues. Distrust of social institutions, including science, increased (see, for example, Brooks 1968). In this climate, the image of the neutral science advisor was shattered. However, at the same time, the institutionalization of science advice increased.

Emblematic of the period were the controversies generated by PSAC under the Nixon administration. The special science advisor (chair of PSAC) had been losing influence in general since the early 1960s, but PSAC had been growing in size until it involved "several hundred prominent scientists and engineers" by 1971 (Perl 1971, 1211). It had been continually expanding the scope of issues it addressed throughout the 1960s. The power of PSAC, however, was to take a sharp nosedive under Richard Nixon after two telling scandals. The first was over the Nixon antiballistic missile (ABM) program. Although official PSAC views on the ABM program were kept confidential, a series of events made clear to the public the doubts PSAC had for the program. In a Senate hearing, former PSAC chairs and members heavily criticized the program, and then current PSAC members refused to sign a letter of support for the program written by the current PSAC chair. Nixon was outraged, and Henry Kissinger "was reinforced in his belief that the committee was filled with administration critics operating under the pretense of scientific neutrality" (Smith 1992, 173). The only scientists who appeared to fully agree with administration views were from the Pentagon. Scientists, "who saw themselves as the essence of objectivity," viewed the episode as "proof that the administration did not want advice but merely public support" (173). Administration officials, on the other hand, considered the behavior of the scientists to be disloyal and believed that science advising "was becoming chaotic" (quoting Kissinger in Smith 1992, 173).

The Super-Sonic Transport (SST) controversy, coming right on the heals of the ABM issue, exacerbated the rift between PSAC and the administration. An ad hoc PSAC panel had been put together to study SST, headed by Richard Garwin. The Garwin Report, finished in 1969, was highly critical of SST's prospects, for both economic and environmental reasons. The report was kept confidential and, in 1970, administration officials then testified in Congress that, "According to existing data and available evidence there is no evidence that SST operations will cause significant adverse effects on our atmosphere or our environment. That is the considered opinion of the scientific authorities who have counseled the government on these matters over the past five years" (quoted in von Hippel and Primack 1972, 1167).

This was in direct contradiction to the Garwin Report, however. When Garwin was asked to testify before Congress on the issue, he made clear that the PSAC report was contrary to the administration's position. Concerted effort by a member of Congress got the report released, and the administration was sorely embarrassed (von Hippel and Primack 1972, 1167). After his reelection in 1972, Nixon asked for and received the resignations of all PSAC members and refused to refill the positions (Smith 1992, 175). He transferred the science advisor position to the National Science Foundation, out of the executive office. PSAC had been killed and the science advisor position would not be reinstated to the executive office until 1976, and then only by an act of Congress (178).

The most obvious lesson of these PSAC controversies was the danger of confidential science advice to the executive branch, particularly when dealing with issues not related to national security. The executive branch could either make public the detailed advisory report or keep it confidential, and in either case, make claims that the advisors supported the project or policy choice. Many observers saw this as a dangerous imbalance of power and/or an abuse of science's authority. Thus Martin Perl lamented on the limited advice scientists gave to Congress compared to the executive branch and criticized the confidentiality of much science advice: "Large numbers of advisory reports are made public; but, unfortunately, it is just those reports which concern the most controversial and the most important technical questions that are often never made public, or only after a long delay. This is unfortunate . . . for the process of making technical decisions in a democracy" (Perl 1971, 1212–13). The problem of confidentiality led Frank von Hippel and Joel Primack to call for "public interest science," in which scientists would act to advise the public, Congress, or special interest groups directly in order to counter the authoritative but secretive weight of executive science advice (von Hippel and Primack 1972).

Congress came up with a legislative response to the problem, however. After a lengthy development, the Federal Advisory Committee Act (FACA) was made law in 1972. Among its provisions, which govern all federal advisory committees, are requirements that advisory committee meetings be publicly announced well prior to the meeting date, that meetings be open to the public (with few exceptions), that meeting minutes be made available to the public, and that committees be balanced in their composition (for more details, see Jasanoff 1990, 46–48; Smith 1992, 25–31). Openness to the public was to help prevent potential abuses of power. Advisory committee reports as well as deliberations were to be public, except when the necessity of national security dictated otherwise. Advice on public issues concerning the environment, consumer safety, or transportation choices could not be secret anymore. The advisory system would no longer be able to be used to create "a facade of prestige which tends to legitimize all technical decisions made by the President" (Perl 1971, 1214).

In addition to dealing with the problem of secret advice, FACA also addressed the second lesson apparent from both the PSAC and other prominent controversies of the day (such as the debates over DDT): scientists can and would enter the political fray, often on opposing sides of an ostensibly technical issue. The disinterested and neutral science advisor was no longer a tenable model, and the requirement of balance in the composition of advisory committees was a congressional response to the demise of that model. It was apparent that scientists could be experts and advocates simultaneously, and so a balance of the range of scientific views was needed on advisory committees, in addition to more traditional concerns with getting the appropriate expertise. (Indeed, von Hippel and Primack called on scientists to be both advocates and experts.) Other observers also rejected the disinterested advisor model. Brooks and Skolnikoff, in a discussion of science advising, explicitly rejected the view that scientists could provide purely "value-free" advising (Skolnikoff and Brooks 1975). And Dorothy Nelkin, in her famous essay "The Political Impact of Technical Expertise," provided a detailed look at how scientists could disagree with each other over technical issues in the political arena, and the political impact/use of such technical disagreements (Nelkin 1975). By 1980, the ideal of the disinterested science advisor was dead. David Bazelon wrote in the pages of *Science*: "We are no longer content to delegate the assessment and response to risk to so-called disinterested scientists. Indeed, the very concept of objectivity embodied in the word disinterested is now discredited" (Bazelon 1979, 277). Instead of neutral advice, the predominant image was one of dueling experts, battling it out along political and scientific fault lines.

Despite the criticisms of science advising and doubts expressed about its role in the early 1970s, the need for scientists in advisory roles continued to expand, particularly with the expansion of the regulatory agencies in that period. Although PSAC was dismantled by the executive office, this singular decline was more than made up for in the expansion at the regulatory level. Even in 1971, estimates of the number of scientists involved in the advisory apparatus ran into the thousands (for example, see Perl 1971, 1212). New legislation (such as the National Environmental Protection Act, Clean Air Act, and Clean Water Act) increased the need for science advice to inform public policy. New agencies were created to implement the regulatory laws (for example, the Environmental Protection Agency, Occupational Safety and Health Administration, and Consumer Product Safety Commission). To provide these agencies with independent (from their staff scientists) advice, science advisory boards were created and proliferated at an astonishing rate.[4]

These boards, while wielding no direct power over agencies (such power would be unconstitutional) have become a major tool for the legitimation of policy decisions. Often created initially by the agency in order to help legitimate their science-based policy decisions, these boards became mandated by Congress in many cases in order to ensure that science used in policymaking would be sound. A negative or highly critical review by an advisory board almost always sends a policy back to the agency for reconsideration.

However, in the 1970s, given the prevailing image of science advisors, these advisory boards did little to squelch technical controversy over policy. While it was hoped that they might decrease the adversarial debates that were becoming the norm in science-based policymaking, they did not. Rather, they became yet another player embroiled in the technical adversarial debates.[5] The scientific uncertainties usually went too deep to be settled by adding simply another layer of scientific review. Even when multiple review panels came to the same conclusions, it is not clear that the scientific issues were settled. (For case studies, see Jasanoff 1990, chap. 2.) In addition, the National Research Council's reports, considered the gold standard for science advice, often failed to produce clear consensus statements on key policy points such as the risk saccharin poses to humans (Bazelon 1979, 278). Some new structure was needed to resolve the endless debates and restore the credibility of science-based decision-making. Redefining the boundary between science and policy seemed to be the answer.

Procedurally Defined Boundary, 1980–2000

By 1980, the intolerability of the dueling experts model was apparent. Dueling experts made the scientific input to the policymaking process appear to be entirely political. A cynic might note that this undermined the ability of policymakers to point to science as a way to bolster their decisions, or to hide their own responsibility for decision making behind the authority of science. At the very least, having contention reach all the way back into the scientific basis for a policy decision greatly slowed the policymaking process and hampered agencies with additional court challenges. The scientific reputation of agencies was also tarnished, thus decreasing their general credibility. The boundary between science and policy needed to be reestablished, but an ontologically defined boundary was no longer tenable. The ideal of scientists as essentially politically neutral had been shattered.

Regulatory agencies had begun struggling with this problem by the mid-1970s, as court cases challenged the scientific basis for their decisions, and they were repeatedly embarrassed by public criticism of their science. Because the controversial cases often involved uncertain risks to the public, the developing field of risk analysis seemed to offer a plausible solution to the boundary problem with a procedurally defined boundary to distinguish the realm of science from the realm of policy.[6] Canonized in the 1983 National Research Council work *Risk Assessment in the Federal Government*, risk analysis divided the decision-making process into two parts, risk assessment and risk management. As defined in the National Research Council text: "Risk assessment is the use of the factual base to define the health effects of exposure of individuals or populations to hazardous materials and situations. Risk management is the process of weighing policy alternatives and selecting the most appropriate regulatory action, integrating the results of risk assessment with engineering data and with social, economic, and political concerns to reach a decision" (NRC 1983, 3).

Risk assessment was to provide a technically based, scientifically sound assessment of the relevant risks and was thus to be the realm of science. Risk management was to consider options to deal with that risk, including the possibility that the risk might be deemed acceptable. The contentious political wrangling was thus to be contained in the risk management part of the process and kept out of the risk assessment.[7]

The risk assessment part was not naively designated as value free, however. The National Research Council recognized that because of incomplete scientific information, some inferences would have to be made in order to bridge the gap between accepted scientific information and an assessment of risk that would be of use to policymakers (NRC 1983, 48). Without such inferences, current scientific information is often unable to say something specific enough to be of help to policymakers.[8] Agency scientists drafting risk assessments must decide which inference to make, and the scientists often need to go beyond widely accepted scientific information to complete the risk assessment. For example, in considering chemical risks to human health, one might need to extrapolate from high experimental doses to low environmental doses. There are a range of plausible extrapolation methods available, including linear extrapolation and threshold models, which can produce a wide range of final numbers (several orders of magnitude apart) in the risk assessment. Rarely does the available scientific information clearly justify a preference for one model over another.

In general, the National Research Council recognized that such inferences "inevitably draw on both scientific and policy considerations" (NRC 1983, 48). However, the National Research Council did not think that the policy considerations specific to the case at hand should have an influence in shaping the inferences risk assessors made. In order to shield agency scientists from the political pressures that arose in risk management, the National Research Council recommended that agencies institute "inference guidelines," defined as "an explicit statement of a predetermined choice among options that arise in inferring human risk from data that are not fully adequate or not drawn directly from human experience" (NRC 1983, 51). Thus, instead of a choice shaped by the particulars of the specific — policy case, the choice should be predetermined by general policy considerations, such as the degree of protectiveness the agency wishes to uphold, as well as accepted scientific considerations, such as what the currently accepted pathways for producing cancer are.

An example will make this point clearer. Regulatory agencies generally wish to be protective of public health and thus will adopt moderately conservative inference guidelines. It is rare that one has strong evidence that substances will cause cancer in humans (direct laboratory testing of humans is morally unacceptable and epidemiological evidence is usually inconclusive). Thus, a standard inference guideline is to assume that if a substance causes cancer in laboratory rodents, it will likely cause cancer in humans and should be considered a likely human carcinogen. Given that most mammals have similar responses to many substances, such an assumption has some scientific basis, but the blanket assumption of similarity also has a strong policy basis of protectiveness of human health. Once one makes this assumption of similarity among mammals, rodent data (derived from high doses) can then be used to extrapolate risks of cancer to humans (usually exposed to low doses), and used to set acceptable dose levels. The actual method of extrapolation involves yet another inference guideline and additional concerns over the degree of protectiveness.

The key, according to the 1983 National Research Council committee, was to make these decisions ahead of time, before considering a particular chemical. The explicit inference guidelines were to assure that the risk assessment process, while not being completely scientific, would not be unduly influenced by the *specific* economic and social considerations of the case at hand, and would thus retain its integrity, based on its stance as free from (the particular) values. As the National Research Council saw it, the maintenance of such integrity was central to the reasons for a strict risk assessment/risk management distinction: "If such case-specific considerations as a substance's economic importance, which are appropriate to risk management, influence the judgments made in the risk assessment process, the integrity of the risk assessment process will be seriously undermined" (NRC 1983, 49). In other words, the point of making a distinction between the two phases was to defend the integrity of the first phase, where presumably science could have its best say on what the risks were. Properly developed inference guidelines could maintain this boundary between the risk assessment and risk management, and provide a clear procedural distinction between the role of science and the role of politics in policymaking.[9]

The National Research Council foresaw the difficulties that have come to haunt inference guidelines. Most obvious and important was (and remains) the problem of rigidity versus flexibility. The more rigid a guideline, the more predictable and transparent its use would be, and the more consistently it would be applied, eliminating the need for judgment and the values that come with judgment. However, such rigid guidelines might run roughshod over scientific complexities of specific cases. To use the example mentioned earlier, there are cases where a type of cancer does appear in chemically exposed laboratory animals, but we have good reason to think that the cancer will not appear in similarly exposed humans. The case of saccharin exemplifies this well: the rats used in the cancer study have a particular substance in their kidneys that reacts with saccharin to produce cancer; humans do not have this substance. With a fuller understanding of the mechanism (and a fairly simple mechanism at that), the inference from rat cancer to human cancer appears unwarranted. An inference guideline that requires the inference from rat cancer to human cancer would ignore important scientific evidence; an inference guideline that considered the additional evidence would require specific judgment on the weight of evidence needed to override the default, thus undermining the advantages of the guideline.

This tension inherent in the inference guidelines has not been resolved. Recent EPA attempts to revise its cancer-risk assessment guidelines have struggled with precisely this issue. The agency has decided largely in favor of a high level of flexibility in the guidelines, requiring much individual and case-specific judgment. This can be seen in the language of default assumptions. Consider the following sample assumption: "The default assumption is that effects seen at the highest dose tested are appropriate for assessment but that it is necessary that the experimental conditions be scrutinized" (EPA 1999, 1–12). This gives wide latitude for the risk assessor to accept or reject experimental findings based on the specific conditions of the testing (for example, whether or not doses are high enough to cause liver cancer exclusively through cell death and replacement proliferation, suggesting that the substance is not *really* a carcinogen). Individual judgment is still needed to weigh the case-specific evidence. In the hazard assessment phase of risk assessment, the need for judgment is even more explicit. The EPA has chosen to pursue "weight-of-evidence narratives," with no formula for how to proceed: "The factors are not scored mechanically by adding pluses and minuses; they are judged in combination" (2–35). Even the descriptors for the hazard (categories that describe likely general risk, such as "carcinogenic to humans" and "likely to be carcinogenic to humans") require judgment to apply: "Applying a descriptor is a matter of judgment and cannot be reduced to a formula" (2–43). While the allowance for case-specific judgment allows one to include the most recent and complex science in one's risk assessment, it also works against the original purpose of the inference guidelines: to reduce individual, case-specific judgment, with their attendant potential for influence by social or political values.

Because of the need for flexibility in inference guidelines, the original purpose of the boundary between risk assessment and risk management has been sorely undermined. If case-specific values are needed to implement the inference guidelines, then case-specific values will be needed in the risk assessment phase of the policy process. If case-specific values are needed in risk assessments, then the boundary that was to keep case-specific values from interacting with science has failed.

This result, the failure of the boundary between risk assessment as the realm of science and risk management as the realm of politics and values, is reflected in the recent disputes over policymaking. Not only have there been contentious battles over the inference guidelines to be instituted by agencies, but there have been continued battles over the science to be used in policymaking. Charges of "junk science" and calls for "sound science" are common in the debates over how to interpret and use in policymaking the science we have available. As a recent analysis by Charles N. Herrick and Dale Jamieson shows, however, "junk science" is usually (over 80 percent of the time, according to their survey) simply science whose implications or emphasis one does not like, rather than science that genuinely does not meet basic methodological strictures (Herrick and Jamieson 2001). "Sound science" is used in these debates as a rhetorical opposite to "junk science." But the suggestion of those who use these terms is that sound science is firmly on the science side of the boundary, and junk science is science contaminated by politics. That the terms are used merely rhetorically most of the time in policy disputes indicates the stress the boundary is under.

A recent article in the *American Journal of Public Health* (January 2002) also exemplifies the current tensions over the boundary. The article, "Attacks on Science: The Risks to Evidence-based Policy," describes what the authors see as a wide array of "tactics used to undermine sound science" (Rosenstock and Lee 2002, 14). While some of the tactics are extremely disturbing, such as the "efforts to squelch unwanted scientific findings" by private interests, others reflect the uncertainties over how to interpret evidence and whether the evidence we have available is adequate to support certain policies (16). These include emphasizing some studies over others, calling for greater peer review of science-based regulatory documents, and court challenges to the scientific bases of regulatory action. The article's authors view these actions as attempts to "exploit scientific uncertainty to deflect attention from what is known and from the actions that would credibly follow that knowledge." As we will see shortly, the problem of scientific uncertainty makes a firm boundary between science and policy untenable; there is no firm bedrock of the known which can guide actions with complete surety. That these problems are seen in terms of boundary issues is made clear by the end of the article, where the authors argue that the most important step that must be made is "to more rigorously define and clarify the boundary between science and policy" (17). Precisely how to do this is not made clear.

That political disputes continue to reach back into the risk assessment phase of policymaking, the purported realm of science free from (at least case-specific) values, makes clear that this most recent attempt at drawing boundaries between science and policy has failed. This should neither surprise nor dismay us, as I will argue. Any attempt to define a boundary between science and politics based on the role of values is doomed to failure. Values, including nonepistemic social values, are a necessary part of scientific reasoning, particularly for science to be used in policymaking. This does not mean that science loses its value to the policymaking process, nor that anarchy will result. In the conclusion I will note recent attempts to redefine the policy process in terms of analysis and deliberation, where neither mode of the process is the exclusive province of science. Such an approach can provide an interesting model for understanding how science and values both need to play a role throughout the policymaking process.

Explaining the Failure of Boundaries Between Science and Policy

Both of the boundaries between science and policy that have been erected since World War II have relied on some view of the role of values in science versus the role of values in policy to hold the two apart. For the ontologically defined boundary, scientists were looked upon as inherently apolitical creatures, capable of functioning above the value-laden political fray. In the procedurally defined boundary, the process of policymaking was defined so that science could provide its input in a neutral space, a part of the process free from the particular values of the case at hand. Both boundaries have failed.

A reevaluation of the role of values in science clarifies why they have failed. As I have argued elsewhere (Douglas 2000), the process of doing a scientific study requires the consideration of a range of values. If the study has clear policy implications, those values must include social and ethical values as well as epistemic values.

Let us briefly review that argument. In the doing of science, whether for use or for pure curiosity, scientists must make choices. They choose a particular methodological approach. They make decisions on how to characterize events for recording as data. They decide how to interpret their results.[10] Scientific papers are usually structured along these lines, with three internal sections packaged between an introduction and a concluding discussion. In the internal sections of the paper (methodology, data, results), scientists rarely explicitly discuss the choices that they make. Instead, they describe what they did, with no mention of alternative paths they might have taken.[11] To discuss the choices that they make would require some justification for those choices, and this is territory many scientists would prefer to avoid in explicit discussion. It is precisely in these choices that values, both epistemic (and/or cognitive) and, more controversially, nonepistemic (social or ethical), play a crucial role. Because scientists do not recognize a legitimate role for social values in science (social values would damage "objectivity"), scientists avoid discussion of the choices that they make.

How do the choices *require* the consideration of epistemic and nonepistemic values? Any choice involves the possibility for error. One may select a methodological approach that is not as sensitive or appropriate for the area of concern as one thinks it is, leading to inaccurate results. One may incorrectly characterize one's data. One may rely on inaccurate background assumptions when interpreting results.[12] In areas where the science is used to make public policy decisions, such errors lead to clear nonepistemic consequences. To weigh the relative seriousness of errors, one would need to assign values to the various likely consequences, both epistemic and nonepistemic. Only with such evaluations of likely error consequences can one decide whether, given the uncertainty and the importance of avoiding particular errors, a decision is truly appropriate. Values are an important, although not determining, factor in the making of internal scientific choices.

If this is true for an individual scientific study, then it is certainly true for the attempts made by scientists to consider large sets of studies in the process of assessing risk and providing policy advice. There are always conflicting studies to be weighed, alternative methodological approaches for the same issue, and different theoretical perspectives from which to choose. All of these acts require scientific judgment. Such judgment, however, is not and should not be — value free (neither epistemic nor nonepistemic). Such judgments need to weigh the strength of evidence *and* consider the impacts of error. Only then is the reasoning complete, and only then are scientists meeting their moral responsibilities to consider the impacts of their choices (Douglas 2003).

One might counter this argument for the necessity of values in science with the claim that if, in fact, value judgments are needed to do policy-useful science, then scientists should not be the ones making those value judgments, both to preserve the value-free nature of science and to preserve the democratic nature of our political processes, where publicly accountable figures make the value judgments. In order for this to work, publicly accountable figures would have to work with scientists throughout their research projects, overseeing the judgments that are made along the way. They would also have to oversee the integrative work of scientists pulling together relevant studies for a comprehensive overview of a subject. Every place scientists make judgments, from methodological choices to interpretations of a set of studies, would require such oversight.

Two obvious problems arise here. The first is purely pragmatic: Where would we find this vast cadre of publicly accountable officials to perform this scientific oversight? The second concerns the need in such a system for cooperation from scientists. If the public oversight is to have any meaning, the overseers must be able to overturn a judgment made by a scientist when the balance of values is deemed inappropriate. Such continual oversight of and interference with their work would be very annoying to scientific researchers, and would constitute a deep undermining of their much valued autonomy. To curb the influence of the overseers and protect some semblance of autonomy, scientists could simply pretend that they are not making any particular judgments, that there is no uncertainty, and that much of their work is standard rote procedure. Thus, the crucial judgments would become even more hidden from public accountability, buried even deeper in the science than they are today. The threat to scientific autonomy and the practical difficulties that would accompany such a system should make us reconsider whether this course of action is worth pursuing to have a clear boundary between science and policy. It would be better both for science and for clarity in policy to acknowledge and accept the necessity of value-laden decisions by scientists doing science.

The burden, the responsibility for thinking through the full implications of scientific work, must fall primarily on the scientist.[13] There are ways to help scientists shoulder this burden. If social agreement can be achieved on which values should be predominant, or on what the proper weighing should be of frequently competing values (such as public health versus industrial productivity), scientists can use that consensus to guide the value judgments that shape their work and interpretations. In areas of high dissensus, however, scientists are left to their own devices. What they must do is make their choices as explicit as possible, as well as the reasoning behind those choices. Only then can the public and policymakers understand the value-laden complexity involved in scientific work. Such understanding would allow people to make better use of the science and to have more accurate expectations for the products of science advice.

Conclusion

If the arguments I make here are correct, science to be used in policymaking *should not* be value free. Any boundary, ontological, procedural, or other, that attempts to divide science and policy on the basis of the presence of values in policy or the absence of values in science is thus doomed to fail. Social values necessarily infect both science and policy. But in both realms, the role of values is not to replace evidence, but to help weigh it. Even in politics, strict clinging to a strongly held view in the face of overwhelming evidence to the contrary is looked down on as dogmatic. Recognizing the proper role for values in scientific reasoning will not produce intellectual anarchy. It does, however, call for a serious reexamination of the relationship between science and policy.

Recent work by the National Research Council has begun to forge such a new picture, with no strict borderlines between science and policy in the policymaking process. Published in 1996, the National Research Council's *Understanding Risk* set forth a new framework for the risk analysis and policymaking process, in terms of analysis and deliberation. Neither of these modes is solely the province of science; scientific processes are understood to contain elements of both analysis and deliberation. The ideas provide the beginnings of a conceptual structure for policymaking, *sans* science policy boundaries. Fleshing out this conceptual structure, in both theory and practice, provides a way out of the fruitless border skirmishes of the past half century.

NOTES

I would like to thank the University of Pittsburgh and the University of Konstanz for creating the impetus for this chapter, and the National Science Foundation (SDEST grant no. 0115258) for the funding, which gave me time to write it. I am also grateful to Sandra Mitchell for her insightful comments on an earlier draft.

1. There was, however, no postwar consensus on how the government would decide which contracts to fund. Bush and his allies argued for a "best science" approach, with scientists controlling the decisions and purse strings, whereas Senator Kilgore and others argued for distributing the funds geographically to foster

science in all states and for an emphasis on science that would serve the public good. These debates delayed the creation of the National Science Foundation until 1950. For a more complete account see Kleinman (1995).

2. Basic research received roughly $200 million from the government in 1956 (Bush 1945/1960, xxv).

3. James Killian became the first formal presidential science advisor, to be followed by George Kistiakowsky in 1959, and then Jerome Wiesner in 1961 (Burger 1980, 8).

4. For a quick overview of the multiple advisory panels created in the 1970s for the Environmental Protection Agency, Food and Drug Administration, Occupational Safety and Health Administration, and Consumer Product Safety Commission, see Jasanoff (1990, 34–35). Later chapters of that same work provide more detailed looks at the history of the advisory committees.

5. Examples of this type of controversy in the 1970s include the regulation of pesticides such as 2,4,5-T (Smith 1992, 78–82), dieldrin, aldrin, heptachlor, and chlordane (see Karch 1977 for these cases), debates over air-quality standards (Jasanoff 1990, 102–12), and the debate over saccharin.

6. For a parallel but different account of the developing focus on process in this period see Jasanoff (1992).

7. The National Research Council argued for conceptual separation between risk assessment and risk management, but not an institutional separation. If one agency did risk assessments and another risk management, the National Research Council warned that the risk assessments produced would likely be useless to actual policymaking.

8. For example, if a substance was known to cause cancer in laboratory animals but there was no data for humans, one would have to make an inference from the available animal data in order to state anything concerning the risk posed to humans by that substance. One might infer that because the laboratory animals studied are mammals, and serve as good models for humans, we should presume the substance is a carcinogen for humans too, and treat it as such for regulatory purposes. With this inference made, determining what doses pose what risks to humans requires additional inferences from the animal data to the human population. On the other hand, one might decide to infer that because the animals studied are not a good model for humans, there is no reason to believe that the substance poses a cancer risk to humans. The laboratory animal studies themselves often do not indicate clearly what the most plausible inference is to draw concerning human risk.

9. William Ruckelshaus, who headed the EPA in the mid-1980s and attempted to institute the National Research Council's recommendations concerning risk assessment, presented two different reasons to keep risk assessment separate from risk management. The first, to keep the science in risk assessment from being adversely affected by political issues, he articulated in 1983 (Ruckelshaus 1983). The second, articulated two years later, was to keep the technical experts from taking over the "democratic" risk-management process. The first reason seems to have had more historical importance, particularly in light of the scandals arising in the Reagan-Gorsuch EPA (1981–1983), and the administration's deregulation policies (see Silbergeld 1991, 102–3).

10. Richard Rudner (1953) made a similar point about the practice of science, although Rudner focused solely on the scientist's choice of a theory as acceptable or unacceptable, a choice placed at the end of the "internal" scientific process.

11. It can be difficult to pinpoint where scientists make choices when reading their published work. One can determine that choices are being made by reading many different studies within a narrow area and seeing that different studies are performed and interpreted differently. With many cross-study comparisons within a field, the fact that alternatives are available, and thus that choices are being made, becomes apparent.

12. Note that in reading a scientific paper with any one of these kinds of errors, it would not be necessarily obvious that a choice, much less an error, had been made.

13. Another way out of this line of argument is to suggest that scientists report findings with error ranges that capture all the uncertainties. However, this does not eliminate the need for judgment, but simply pushes it back to the level of error estimation. The scientist must then make a judgment about the error rates for any given choice he or she makes, including methodological choices. A judgment requiring values (such as, Is this error rate accurate enough?) is still being made. To try to push the values back further only creates a regress and compounding error rates, leading to greater confusion. For an early version of this argument see Rudner (1953).

In Douglas (2003) I present a more thorough and complete argument for why thinking about the consequences of scientific error is an unavoidable moral responsibility of scientists.

REFERENCES

Bazelon, David. 1979. Risk and responsibility. *Science* 205:277–80.
Brooks, Harvey. 1964. The Scientific advisor. In *Scientists and national policymaking,* ed. Robert Gilpin and Christopher Wright, 73–96. New York: Columbia University Press.
———. 1968. Physics and the polity. *Science* 160:396–400.
———. 1975. Expertise and politics: Problems and tensions. *Proceedings of the American Philosophical Society* 119 (4):257–61.
Burger, Edward. 1980. *Science at the White House: A political liability.* Baltimore, MD: Johns Hopkins University Press.
Bush, Vannevar. 1945/1960. *Science: The endless frontier.* Repr. National Science Foundation, Washington, DC.
Douglas, Heather. 2000. Inductive risk and values in science. *Philosophy of Science* 67:559–79.
———. 2003. The moral responsibilities of scientists: Tensions between autonomy and responsibility. *American Philosophical Quarterly* 40 (1):59–68.
Dupre, Hunter. 1957. *Science in the federal government: A history of policies and activities to 1940.* Cambridge, MA: Harvard University Press.

Environmental Protection Agency. 1999. Draft guidelines for carcinogen risk assessment. Risk Assessment Forum, NCEA-F-0644.

Gilpin, Robert. 1964. Introduction: Natural scientists in policymaking. In *Scientists and national policymaking*, ed. Robert Gilpin and Christopher Wright, 1–18. New York: Columbia University Press.

Gilpin, Robert, and Christopher Wright. 1964. *Scientists and national policymaking*. New York: Columbia University Press.

Greenberg, Daniel. 1967/1999. *The politics of pure science*. Chicago, IL: University of Chicago Press.

Herrick, Charles N., and Dale Jamieson. 2001. Junk science and environmental policy: Obscuring public debate with misleading discourse. *Philosophy and Public Policy Quarterly* 21 (2/3):11–16.

Hewlett, Richard, and Francis Duncan. 1962/1990. *Atomic shield*. Repr. Berkeley: University of California Press.

Jasanoff, Shiela. 1990. *The fifth branch: Science advisors as policymakers*. Cambridge, MA: Harvard University Press.

———. 1992. Science, politics, and the renegotiation of expertise at EPA. *Osiris* 7:195–217.

Karch, Nathan. 1977. Explicit criteria and principles for identifying carcinogens: A focus of controversy at the Environmental Protection Agency. In *Decision making in the Environmental Protection Agency*, vol. 2a, *Case studies*, 119–206. Washington, DC: National Academy Press.

Kevles, Daniel. 1971/1995. *The physicists*. Cambridge, MA: Harvard University Press. Page references are to 1995 edition.

Kleinman, Daniel Lee. 1995. *Politics on the endless frontier: Postwar research policy in the United States*. Durham, NC: Duke University Press.

National Research Council [NRC]. 1983. *Risk assessment in the federal government: Managing the process*. Washington, DC: National Academy Press.

———. 1996. *Understanding risk: Informing decisions in a democratic society*. Washington, DC: National Academy Press.

Nelkin, Dorothy. 1975. The political impact of technical expertise. *Social Studies of Science* 5:35–54.

Perl, Martin. 1971. The scientific advisory system: Some observations. *Science* 173:1211–15.

Pursell, Carroll W., Jr. 1965. The anatomy of failure: The science advisory board, 1933–1935. *Proceedings of the American Philosophical Society* 109:6.

Rosenstock, Linda, and Lore Jackson Lee. 2002. Attacks on science: The risks to evidence-based policy. *American Journal of Public Health* 92 (1):14–18.

Ruckelshaus, William. 1983. Risk, science, and public policy. *Science* 221:1026–28.

Rudner, Richard. 1953. The scientist *qua* scientist makes value judgments. *Philosophy of Science* 20:1–6.

Silbergeld, Ellen. 1991. Risk assessment and risk management: An uneasy divorce. In *Acceptable evidence: Science and values in risk management*, ed. Deborah Mayo and Rachelle Hollander, 99–114. New York: Oxford University Press.

Skolnikoff, Eugene, and Harvey Brooks. 1975. Science advice in the White House? Continuation of the debate. *Science* 187:35–41.

Smith, Bruce. 1992. *The advisors: Scientists in the policy process.* Washington, DC: Brookings Institution.

von Hippel, Frank, and Joel Primack. 1972. Public interest science. *Science* 177: 1166–71.

Wood, Robert. 1964. Scientists and politics: The rise of an apolitical elite. In *Scientists and national policymaking*, ed. Robert Gilpin and Christopher Wright, 41–72. New York: Columbia University Press.

12

The Prescribed and Proscribed Values in Science Policy

Sandra D. Mitchell
Department of History and Philosophy of Science, University of Pittsburgh

In a series of articles Heather Douglas investigates the role of values in science, especially when that science is advising policy, and the ways in which normative considerations must direct the judgments of scientists (Douglas 2000, 2003; chap. 11, this volume). In her chapter in this volume (chap. 11) Douglas challenges the assumption that there can be a sharp boundary separating science and policy even when this boundary has been posited to protect science from political corruption and locate political responsibility in the hands of elected or appointed public servants. Her strategy is to document the history of governmental structures for ensuring that the two sets of practices stay in their proper place. From a history of shifting roles, Douglas hopes to draw both descriptive and normative conclusions: that keeping the two sides of the divide in separate territories is impractical and that trying to do so is undesirable.

Douglas divides the history of U.S. government/science organizational structures into three periods marked by varying roles for values in science. In the first period, 1945–1965, science was seen to be socially beneficial and a number of policy advisory structures were initiated. At the same time, science was deemed "free from social and political values." In the second period, 1965–1980, values in science for policy became apparent through a couple of scandals in the Nixon era, and institutional remedies to remove the influence of values were attempted. In the third period, 1980–2000, values were again detected in science, this time through the failure of advisory committees to generate consensus positions, but the institutional remedy incorporated a view that values were not removable, only that they should be contained in a principled way.

Douglas's evidence for these shifting conceptions of the role of values in science comes, in part, from how governmental structures responded to perceived "problems." Two cases examined are the failures of government to listen to science advice regarding antiballistic missiles (ABM) and supersonic transport (SST) (in the second period) and the failure of science advisory boards to provide consensus views of scientific issues relevant to policy (the third period).

I will examine the logic of the arguments Douglas makes on the basis of her historical sketch and will suggest that there is possible equivocation on the meaning of *values* that carries so much explanatory weight in her argument. The overall structure of the argument is abductive. What account of the role of values in science *best* explains the history of science/policy skirmishes? Her answer: "Science to be used in policymaking *should not* be value free. Any boundary, ontological, procedural, or other, that attempts to divide science and policy on the basis of the presence of values in policy or the absence of values in science is thus doomed to fail" (chap. 11, this volume). This type of historical argument can be challenged, of course, by asking whether that *really* is the best explanation for what happened. I will suggest an alternative interpretation that better captures at least some of the policy texts. I do think Douglas has made the case that her account can explain some features of the history while having to explain away some others. It is certainly a possible account of why what happened happened. The real work, however, is in the thesis itself, which she has little time to defend in her chapter in this volume, but which she does defend elsewhere (see Douglas 2000), to which I will look in trying to clarify the meaning of epistemic and nonepistemic values and the roles they might play in scientific evaluations. Though agreeing with much of Douglas's analysis of the uncertainty that plagues all stages of scientific practice, I draw a different conclusion. Rather than giving up locating a boundary between science and policy on the basis of values, I suggest that it is extremely important to analytically separate, and institutionally recognize, the differences between science and policy. This requires a more detailed account of the types of values that enter into judgments and the contexts of judgment than afforded by the typical dichotomies drawn between epistemic versus nonepistemic values and science versus policy.

The Scandal Case and Its Implications

Douglas recounts the scandal visited on the Nixon administration in the late 1960s as a consequence of its requesting support for its ABM policy from the Presidential Science Advisory Committee (PSAC) and finding that committee members refused to sign a letter of support that the PSAC chair had written. Some of the uncooperative science advisors went even further: they testified before Congress and published their arguments for the technological ineffectiveness of the ABM in *Scientific American*. Nevertheless, scientists at the Pentagon supported the administration's policy. The second scandal was the report of an ad hoc committee of PSAC on SST. The committee was critical of SST for economic and environmental reasons, but the government nevertheless broadcast the view that the panel found no evidence of environmental harm, directly contradicting the scientists' report.

What are these historical episodes supposed to tell us about the role of values in science? According to Douglas, there were two, one that "openness to the public was to help prevent potential abuses of power" and the other that "scientists can and would enter the political fray, often on opposing sides of an ostensibly technical issue"(Douglas, chap. 11, this volume). It is only the first consequence that is clearly supported by the "solution" to the ABM and SST controversies. In their wake Congress passed the Federal Advisory Committee Act of 1972, which made the meetings and minutes of advisory committees public. The act also states that

Any . . . legislation (to establish an advisory committee) shall
  1. contain a clearly defined purpose for the advisory committee;
  2. require the membership of the advisory committee to be fairly balanced in terms of the points of view represented and the functions to be performed by the advisory committee;
  3. contain appropriate provisions to assure that the advice and recommendations of the advisory committee will not be inappropriately influenced by the appointing authority or by any special interest, but will instead be the result of the advisory committee's *independent judgment*. [emphasis added]

At least in this response, the perceived problem was not that the scientists were promoting a specific political agenda that was corrupting their scientific analysis (Douglas's second conclusion), but that political interests or the government itself, by not listening to scientists, could illegitimately influence advisory board members. The insistence on replacing committee confidentiality with disclosure and the attempt to protect "independent judgment" highlights the political character of the executive branch of the government rather than the political character of the scientific assessment in these two scandals.

The question of expert disagreement, however, opens a door that for Douglas lets nonepistemic values into the sanctuary of scientific analysis. Indeed, for her it is the possibility of differing conclusions resulting from judgment based on epistemic values alone that ensures an essential role for nonepistemic values in scientific reasoning. This argument appears to be impeccable. Indeed, under certain interpretations of "values" and "judgment" it is tautological. It would be a mistake to argue with the deductive version of the argument. However, I want to suggest that the there is another interpretation of the nature of the values that fill the gap created by epistemic uncertainty, an interpretation that does not warrant Douglas's conclusion. Furthermore, these other values are not necessarily the ones that the science/policy boundary is aiming to barricade out. In what follows I will lay out different possible responses to the argument from epistemic underdetermination.

The Fact of Scientific Disagreement and the Nature of Scientific Disagreement

Douglas points out that disagreement among scientists has fueled some criticism of scientific objectivity. But the *fact* of disagreement about, say, whether ABM is effective or dioxin causes cancer in humans, by itself does not show that the *source* of disagreement is necessarily nonepistemic values.

When *would* the necessary intervention of nonepistemic values be the right conclusion to draw? The answer is that this conclusion is viable only when we limit what we take to be epistemic values to those that refer narrowly to evidential support alone and when scientific judgment is about the truth or falsity, rather than warranted acceptability, of a causal hypothesis. Narrow epistemic values may include variety of evidence, accuracy of measurement, and replication of experimental results. It is commonplace that evidence, no matter how varied, accurate, or convincing, underdetermines the truth of general causal hypotheses. At the end of the day if a scientist endorses a claim that SST is environmentally harmful or that dioxin causes cancer, the truth of this claim cannot be *guaranteed* by empirical evidence.

However, epistemic values include more than just evidential support. Indeed, ever since the work of Thomas Kuhn (1977), other norms have been recognized as constitutive of scientific judgment, including accuracy of prediction, problem-solving ability, simplicity, and scope (see also McMullin 1983). While evidence underdetermines the truth of a causal claim, the other epistemic values (in McMullin's terminology) or cognitive values (in Laudan's 1984 terminology) can be harnessed to generate a judgment of acceptance. The engagement of any and all these values, though producing reasoned judgment, would never guarantee the truth or falsity of a claim. Inductive risk will always be a danger if truth is what we are after. In addition, employing these additional norms does not *uniquely* determine even the acceptability of a claim. As Kuhn pointed out early on, individuals might prioritize different members of this set of values or interpret their application in different ways (Kuhn 1977). Nevertheless, if the scientists in the advisory committees to policymakers make judgments employing only these broadly epistemic or cognitive values, then there is no necessity for values outside this set to enter the process.

Which Values Settle the Score for Inductive Risk?

Douglas disagrees with the claim stated above. This is the central point of her analysis of inductive risk (Douglas 2000).[1] There she argues that, "When nonepistemic consequences follow from error, nonepistemic values are essential for deciding which inductive risks we should accept, or which choice we should make" (565). Given the underdetermination of the truth of a scientific claim by our epistemic values, if we venture to *accept as true* a claim on the basis of these values, then we could be wrong: it may turn out to be false. Similarly, if we *reject as false* a claim on the basis of these values, we again could be wrong: it may turn out to be true. The risk comes in when we step outside what we can be sure about—such as the degree of support, whether the evidence is supporting or not—to that which we can never be sure about: the truth or falsity of a causal generalization. The culprit in the inductive risk story is underdetermination and the crime is being wrong about the truth or falsity of a causal claim on the basis of empirical support.

Carl Hempel (1965) argued that with inductive risk, as with any risk assessment, the consequences of the different outcomes must be taken into consideration in generating an adequate decision rule. In the case of the possibility of epistemic failure, the values associated with the consequences of such failure are then legitimately involved in deciding to accept as true (or reject as false) a given hypothesis. Douglas expands Hempel's view by arguing that when the consequences of being wrong about a scientific hypotheses are themselves nonepistemic, then nonepistemic values are "essential" or "necessary" for deciding which hypothesis to accept or reject (Douglas 2000, 565).

This expansion to nonepistemic consequences, however, contradicts Hempel's own moral for the inductive risk story. For him, the rules of acceptance would specify only the evidential conditions under which a hypothesis should be accepted, that is, how strong the evidence needs to be to warrant scientific assent. The values that enter in to fill the uncertainty gap are epistemic, those that lead to the goal of "increasingly reliable, extensive, and theoretically systematized body of information" (Hempel 1965, 93). He admits that in practical contexts, that is, when the hypothesis is to form the basis of action or policy, then monetary value or other social values are appropriate to determining acceptability. Douglas justifies her "expansion" of Hempel's argument from inductive risk into the domain of deciding which theory to accept as true by appealing to the authority of science in our society. She claims that, "Where science is 'useful' it will have effects beyond the development of a body of knowledge. . . . If a scientist affirms something as true, or accepts a certain theory, that statement is taken as authoritative and will have effects, potentially damaging ones, if the scientist is wrong" (Douglas 2000, 563). This conflation of the domains of belief and action confuses rather than clarifies the appropriate role of values in scientific practice. Indeed, to make public one's belief that a given hypothesis is true is an action, and in certain contexts a scientist might judge that stating what he or she is scientifically warranted to believe is politically inadvisable. For example, when there was enormous ignorance and misunderstanding about the nature of AIDS, public announcements to the effect that a certain individual had acquired this disease might have been unjustified, even if the belief was fully warranted that the individual did in fact have the disease. The values appropriate to generating *the belief* and the values appropriate to generating *the action* are different. Indeed, one can retain a critical stance with respect to which values are legitimately invoked in a given domain only when they are analytically separated. That a particular individual may be engaged in practices in both belief generation and policy decisions makes it appear that they are inseparable, but blurring the boundary makes criticism of political values entering into the determination of what is or is not a scientifically warranted claim impossible.

Which Values Are Evidenced in the History of Science Policy?

To make reasoned choices in evaluating and accepting a hypothesis, epistemic values, broadly speaking, must play a role in determining those choices. In the context of science policy, however, the issue at hand is which actions or policies should be adopted on the basis of the beliefs that are warranted by those epistemic values. When scientists are involved in policymaking, the very same individual might invoke both sets of values. Analytically, it is crucial to keep the two types of values distinct although they may be embodied in a single individual who is occupying distinguishable roles when he or she participates in a policy committee. One role is that of a scientist and as such the goals that determine the appropriate values that enter into judgments are those that secure warranted belief, for example, replicated experiments and large data sets. The other role, sometimes contractually acquired, is that of a governmental advisor in which the appropriate goals of government, often contested goals, determine the appropriate values that enter into judgments of warranted advice. Indeed, it was my claim earlier that the Federal Advisory Committee Act of 1972 is best understood as bringing advisory committee practices into line with some of the stated goals of government, namely, balancing power among the different branches of government by prohibiting the executive branch from having exclusive access to advisory decisions.

My thesis is that the norms that should govern the actors in science policy are multiple and can be distinguished in terms of "role obligations." Some of the conflict among individuals in deciding policy based on science (or even experienced within an individual, as in the case of moral conflict) is helpfully represented by appealing to the different social institutions and their associated obligations coming into play in generating a judgment. I will appeal to Michael Hardimon's claim that "a role obligation is a moral requirement, which attaches to an institutional role, whose content is fixed by the function of the role, and whose normative force flows from the role" (Hardimon 1994, 334). Determining the one or many institutions relevant to evaluating a judgment or action will help locate which values are "problematic" and which are contributing in a legitimate way.

When looking at judgments in "pure" science, some kinds of nonepistemic values are problematic. That racism should guide the inclusion or exclusion of different data sets is to be condemned (Gould 1981). This is not because racism is contemptible — that must be argued on separate grounds — but because racism does not promote the epistemic goals of science. The intervention of specific values is deemed problematic *because* they counter the constitutive values of the institutions in which the practice is embedded.

In policy contexts both epistemic and nonepistemic values contribute to judgment. This is because two sets of role obligations are involved and not because the two types of values are indistinguishable nor that the two types of values are warranted in each of the individual roles of scientist and policy advisor. Consider the decision as to which of several possible hypotheses supported by agreed upon evidence should be acted on. What values do and should play a role in generating a judgment here? Two social institutions intersect: science and governmental policy. In the science aspect of judgment, the overriding constitutive goal that provides primary obligations is something like that specified by Hempel, namely, reliable information about the world. Any judgment that invokes a value known to counter the obligation to this goal would be deemed illegitimate. In the policy context, what are the constitutive goals that determine the obligations to which a policymaker must adhere to meet his or her role obligation? Policymaking is under the strictures of governmental organizations, and so it is clear that independence from one of the three branches of the government (the executive branch, for example) is constitutive in the U.S. political structure. In addition, in a representational democracy, policy requires representing an array of constituents and stakeholders. Indeed, this would explain the language of the Federal Advisory Committee Act of 1972 that both made the meetings and minutes of advisory committees public and balanced the points of views represented.

With this model we can revisit the issue of conflict among scientists within a policy advisory committee. Does conflict always indicate the intervention of nonepistemic values? The answer is no. Sometimes the individuals in conflict have different interpretations of the ways in which the epistemic values are involved in warranting a scientific judgment or the acceptability of a specific hypothesis. For different individuals, having larger samples may weigh more or less significantly in comparison to measuring quantities more precisely. Alternatively, the lack of consensus may stem from role conflict among and within individual participants. With respect to the distinction between private and public roles and their corresponding obligations, W. T. Jones has argued that "people who view a given role as primarily public and people who view that role as primarily private will disagree about what good performance in that role is and hence about what one's duty is" (Jones 1984, 603–4). In the context of science policy, the same judgment might apply to the question of what are the appropriate norms to consider when viewing one's role on an advisory committee as a scientist and viewing one's role as a defender of the goals of a just society. Indeed, by partitioning obligations by means of distinguishable social institutions, one can begin to understand the personal conflict scientists face when attempting to act respectably in two contexts that require different behaviors. What counts as sufficient evidence for scientific assertability is generally much higher than what policymakers require for action. I believe this is as an instance of the more general problem of differentiating the standards of assertability demanded in the domain of action (even when we do not have perfect knowledge, we have to do something now) and those demanded in the domain of theory (where remaining silent until sufficient evidence is acquired does not have the same costs). The ambiguity of the standards of assertability between science and policy can only be untangled if the normative constraints operating in the different social intuitions are kept separate.

Conclusion

Scientists can certainly disagree about the risk of a potential carcinogen when they disagree about the significance of different epistemic values or the degree of their satisfaction — without nonepistemic values entering in. So why should we think that the kinds of values that scientists reason from in the process of coming to accept an underdetermined theory as true are necessarily nonepistemic? For Douglas, "when nonepistemic consequences follow from error, nonepistemic values are essential for deciding which inductive risks we should accept, or which choice we should make" (Douglas 2000, 565).

But this does not yet point clearly to the values that legitimately play a role in science policy and those that should be eschewed. Surely nonepistemic consequences can *always* follow, even for the most politically innocuous theories. Individual fame, fortune, possibility of promotion, and so on are at risk when a scientist declares the truth of a hypothesis, since it can turn out later to be demonstrably false. But the values relevant to Douglas's historical narrative are political values, *not* idiosyncratic personal values, and the values that are required for the inductive risk argument are merely those that are not *narrowly* epistemic, which does not necessarily mean political values. As Douglas claims, in the third period, 1980–2000, the "dueling experts" model was to be replaced, since "the ideal of scientists as essentially politically neutral had been shattered." (Douglas, chap. 11, this volume).

In fact the National Research Council's 1983 study claimed that "the basic problem in risk assessment is the sparseness and uncertainty of the scientific knowledge of the health hazards addressed, and this problem has no ready solution. The field has been developing rapidly, and the greatest improvements in risk assessment result from the acquisition of more and better data, which decreases the need to rely on inference and informed judgment to bridge gaps in knowledge."

The governmental structures seem to me to be an attempt to ensure that cultural, social, and political values do not taint scientists' decisions concerning the projected consequences of the technologies and substances they were studying. That is, the structures serve to locate the appropriate obligations to role in a context where there is both the possibility of role confusion and role conflict. What Douglas can justifiably argue is that some nonnarrowly epistemic values are necessary to generating a rational decision to accept as true a causal claim. But the kinds of values that fill the gap left by underdetermination are not clearly the kinds of values that the science policy institutional structures were aiming to contain.

NOTES

1. Let me add here that Douglas provides compelling arguments in Douglas (2000) that underdetermined methodological choices occur in science in places

other than the final acceptance or rejection of the truth of a causal hypothesis. These choices come in to play as well in what she characterizes as internal stages of science: "choice of methodology, gathering, and characterization of the data, and interpretation of the data" (565). I believe I could ring the same changes at any of these places, so for brevity let me limit myself to the decision to accept or reject a hypothesis.

REFERENCES

Douglas, Heather. 2000. Inductive risk and values in science. *Philosophy of Science* 67:559–79.
———. 2003. The moral responsibilities of scientists: Tensions between autonomy and responsibility. *American Philosophical Quarterly* 40 (1):59–68.
Federal Advisory Committee Act of 1972. Public Law 92–463. Available at http://www.fda.gov/opacom/laws/fedadvca.htm.
Gould, Steven Jay. 1981. *The mismeasure of man.* New York: W.W. Norton.
Hardimon, Michael O. 1994. Role obligations. *Journal of Philosophy* 91:333–63.
Hempel, C. G. 1985. Science and human values. In *Aspects of scientific explanation,* ed. C. G. Hempel, 18–96. New York: Free Press.
Jones, W. T. 1984. Public roles, private roles, and differential moral assessments of role performance. *Ethics* 94:603–20.
Kuhn, Thomas. 1977. Objectivity, value judgment, and theory choice. In *The essential tension: Selected Studies in scientific tradition and change,* 320–39. Chicago: University of Chicago Press.
Laudan, Larry. 1984. *Science and values.* Berkeley: University of California Press.
McMullin, Ernan. 1983. Values in science. In *PSA 1982*, vol. 2, ed. P. D. Asquith and T. Nickles, 3–28. East Lansing, MI: Philosophy of Science Association.
National Resource Council. 1993. *Issues in risk assessment,* Washington, DC: National Academy Press. Available at http://books.nap.edu/books/0309047862/html/index.html.

13

Bioethics

Its Foundation and Application in Political Decision Making

Felix Thiele
Europaeische Akademie, Bad Neuenahr-Ahrweiler

Questions concerning moral problems stemming from research in the life sciences and concerning the adequate methods and instruments for solving them are timely and urgent—especially in light of intense debates on the acceptability of research on human embryonic stem cells, human cloning, and pre-implantation diagnostics, to name only a few current applications of research in the life sciences.

What role does the bioethicist play in dealing with these moral problems in public debates and the process of decision making in science and technology policy? Although it seems that in everyday life making a moral statement is connected with a claim of interpersonal validity, many have argued that bioethics is no more than an exchange of emotions and beliefs of those concerned with or affected by the moral problems at issue. In this view bioethics may have some effect only as complement to or substitution for psychological and pastoral care, but not as a form of argumentation where the opponent may be convinced by using reasons.

Nonetheless "bioethics" is deemed important by many politicians and other participants in public debates, so that the establishment of bioethics committees seems to be seen as a solution—and unfortunately not as a tool to facilitate developing solutions—for the moral problems at issue. But instead of functioning as valuable and fruitful advisory boards, these committees run the risk of being misused as a fashionable (and soon forgotten) weapon in the battle for political opinion leadership.

Some twenty-five years ago Paul Lorenzen and Oswald Schwemmer wrote that ethics as opposed to logic and the philosophy of science "has remained a reserve of un-scrutinized and partly un-scrutinizable talking back and forth." Sadly, this description still applies to the greater part of *bio*ethics. It is therefore still up-to-date when the authors continue that it "might be especially difficult, to prompt the reader to participate in the constructive building of an ethics" — at least when the subdiscipline bioethics is concerned (Lorenzen and Schwemmer 1975, 149).

Since there are so many conflicts spawned by research in the life sciences and since these conflicts are usually termed moral problems, it might nonetheless be worthwhile not to be content with idle discussion, but to suggest a method of bioethics detailing criteria that would allow one to actually determine whether bioethics is successful in mastering moral conflicts.

Some Remarks on the Foundation of Bioethics

Morals and Ethics

The objective of bioethics should be to consult those groups affected by the moral problems generated by medicine and the life sciences, such as medical doctors, patients, scientists, politicians, and the general public. In the course of this consulting process bioethics should suggest recommendations for special cases, as well as propose general regulations and moral standardizations or normatizations of options to act.

In developing recommendations, bioethics — just as ethics as a whole — does not have for this task a special source of knowledge at its disposal, a source of cognition that would allow the proclamation of binding commandments or prohibitions. In short, an ethicist is not someone whose job is to exclude the ordinary mortal from moral discourse, to tell the person what he or she shall do as a result of the ethicist's esoteric thinking.

Rather the bioethicist will fall back on established agglomerates of moral convictions, norms, and recommendations, will critically test and if necessary revise theses agglomerates, or parts of them, with the help of criteria developed in the tradition of philosophical ethics.[1] (The criteria themselves are open for revision, too.) In this view every component used in an argument may be revised if doing so helps to master the moral problem. The only exception is the principle that "conflicts should be mastered using arguments, not brute force." Not adopting or changing that norm is equivalent to dissolving the very project of ethics (discussed in a later section).

In public debates on bioethics one is often confronted with participants claiming that this or that norm is "obviously correct" and therefore "not debatable." This claim contradict what was said in the preceding paragraph. It turns out, however, that this everyday or folk-belief in "indubitable norms" is relatively easy to challenge with examples from the history of morals, such as changes in public morals concerning homosexuality or the once "divine" disease epilepsy. More complicated is the case of theological ethics. Though most theological ethicists presenting their views in the public usually claim to be of the same breed as philosophical ethicists, it seems that the line of divide is exactly constituted by the fact that philosophical ethicists do not adopt a set of "indubitable norms," whereas theological ethicists do.[2]

The terms *morals* and *ethics* are frequently used synonymously in bioethical debates.[3] Because of this, however, one loses the important distinction between on the one hand a grown agglomerate of moral convictions, norms, and recommendations to act that is subject to multiple changes and on the other the method that is concerned with the critical examination of those agglomerates.

Everyone has a set of morals (no matter how this might be disposed), and for any group one can collect or reconstruct the agglomerate of convictions, norms, and recommendations that constitutes the group's binding set of morals. According to this understanding, there does not exist *the* set of morals but a great variety of morals,[4] such as professional morals that are thought to be binding only on certain (professional) groups (the professional ethos like the medical ethos) or religious morals that ought to regulate social intercourse between the members of a religious group as opposed to those in force between external persons.

Morals are in many cases — and concerning the everyday need for decision making probably in most cases — a sufficient basis for questions on what to do and what to omit. Beyond the routine of decision making trained by morals, however, there are numerous cases (heaped in certain working areas) where morals do not deliver satisfying "recommendations to act."[5] The rule of action that a doctor must always act in favor of his patient's well-being, for example, may well be accepted widely. But does a pediatrician act in favor of his patient's well-being when he performs life-sustaining measures on a severely disabled newborn? Questions like this demonstrate the limits of the efficiency of morals yet again.

Bioethics as a Tool for Mastering Conflicts

In the last several decades a variety of bioethical approaches have been developed, most traceable to the diverse classical ethical approaches.[6] The plurality of competing ethical approaches is often used as an argument against the benefit of moral reasoning for practical questions. That this plurality will probably not be resolved should lead to modesty, not resignation concerning the power of moral argumentation, since we should not demand in the field of bioethics what is not to be had in other practical sciences. For example, in applying laws a judge is always left with a degree of discretion within which he or she has to carefully examine the special circumstances of any given case. An economist, too, is left with a degree of discretion when applying a certain economic theory for, say, a forecast of economic growth for the world economy. It seems rather common that as a result of using different economic models the first economist predicts 1 percent growth, the second 2 percent. In the end the actual economic growth may be no more than 0.5 percent. But this "ridiculously low" difference of 200 percent between prediction and outcome has not yet resulted in the claim that the plurality of economic theories is an argument against the benefit of economic reasoning for practical questions. So why should the same type of argument be used against the use of bioethics for practical questions?

One might even go a step further and compare bioethics to the "hard sciences." In the same way that a controversial law of nature may be applied, it is possible to apply a controversial bioethical norm to a moral problem. Though this may seem unsatisfactory on first sight, this situation is not interestingly different from other scientific disciplines: in view of an open question in the natural sciences a working hypothesis may be proposed, in order to test it empirically. In the same way a bioethical theory that is not yet accepted may be applied to a moral problem in order to test the conflict-mastering potential of this theory. (Compare a civic engineer building bridges and a bioethicist: Both pursue certain practical aims, building bridges and mastering conflicts, respectively. For both there are criteria of success, such as that the bridge lasts at least thirty years or that the conflict is mastered.)

Though this plurality of ethical approaches remains a fact, a closer look shows that there is a definite exception to this plurality. The participants in the bioethical debate have one goal in common: *the wish to master conflicts with the help of moral arguing.*[7] Since this is (at best a true) empirical observation, it would be a naturalistic move to turn this observation into a foundation for ethics. It is appealing, however, to place this line of argument on a broader anthropological basis: since for their very survival humans depend on cooperation, it is necessary to develop rules for this cooperation that certainly will include rules for mastering conflicts, since otherwise a self destructive battle would endanger the cooperation as such.[8] Arguing from anthropological constants still runs the risk of naturalism, and I will not pursue this line of thought further, since it is not essential for my argument. It is sufficient to point out that everyone participating in the debate of moral problems in the life sciences subscribes to the view that these problems should be solved nonviolently, so that the task of ethics can be phrased in the following way: *Ethics develops principles for the mastering of moral conflicts in an argumentative manner. Those principles should be applicable independent of the particular circumstances of the situation in which the conflict arises.*[9]

A moral conflict arises when two actors perform actions or pursue goals that are incompatible. Mastering conflicts is not restricted to the solution of conflicts that are already existent; the avoidance of future but foreseeable conflicts is also of importance, especially when bioethics ought to be used as a tool for the active shaping of science and technology policy.

In order to successfully master conflicts independent of the particular situation in which the conflict arises it is in principle sufficient to recommend rules of action that are justifiable not to everyone but only to the (potential) conflict partners. Identifying conflict partners is, therefore, an essential part of mastering moral conflicts, such as the debate on the moral status of human embryos and primates. The problems that are object to professional bioethical investigation involve as a rule a greater number of conflict partners. Each party claims that the moral the party has adopted is or should be binding for all the others parties, too — with the result that the bioethicist is confronted with a multiplicity of moral convictions, norms, and recommendations to act that are (partly) incompatible with one another. In a certain sense mastering a conflict in an invariant manner relative to the involved parties may be expressed as developing universally valid moral norms and norm systems.[10]

Whether this conception of ethics consists in a narrowing of the traditional purposes of ethics remains open in this chapter. It may, however, be possible to reformulate most of the talk about "the good," "values," "dignity," "freedom," and the like as suggestions for an adequate mastering of conflicts caused by the malfunctioning of human cooperation.[11]

Ethics has been introduced as a tool for mastering conflicts. What are lacking so far are criteria that would allow the determination of whether this task of mastering conflicts is *successfully* accomplished in the context of a specific moral conflict. Before developing such criteria, it will be worthwhile to consider briefly the epistemic status of ethical knowledge and the place of ethics in the canon of (scientific) disciplines.

Ethical Knowledge as Justified Belief

In everyday life it seems that a speaker who is making a moral statement is also making a claim of interpersonal validity. In further analyzing this, some authors have drawn an analogy between empirical statements that are interpreted as having a truth value and moral statements. According to this view moral statements have a truth value, or, in other words, there are *moral facts*. A second, widely adopted thesis is that if and only if statements have a truth value, it is possible to determine their validity by way of argumentation. From these two theses it can be concluded that moral statements can be subject to a reasoned argumentation for and against their validity.

Two ethical approaches have been developed on the basis of the two theses: The first position claims that the truth value of moral facts is empirically determinable (descriptivism); the second position denies this and claims instead that moral facts are of a peculiar type that is open to an equally peculiar determination (intuitionism).[12] It is known that both variants are burdened with grave philosophical problems, so that either one of the two theses or the claim that the validity of moral statements can be determined by reasoned argumentation has to be given up. Many if not most scholars in the tradition of analytical philosophy were inclined toward the second option, so now the possibility of proper moral argumentation is held unachievable.[13]

However, the method of moral argumentation has found new interest partly owing to developments in the life sciences, especially medicine (dialysis, contraception and abortion, artificial respiration) and changes in the Western world (particularly the decline of traditional institutions such as the church and the family). It has been convincingly argued that even if moral statements do not have a truth value, they nonetheless can be part of a reasoned argumentation.[14]

This chapter is not the place to exposit the arguments in favor of such a view. It should be pointed out, however, that the view that moral argumentation cannot be reasoned is plainly counterintuitive if one turns to legal argumentation that is part of, or at least very similar to, moral argumentation. In legal argumentation legal norms are applied by using specific rules. Both the rules and the norms are open to debate: rules for drawing conclusions, ways of justifying the norms used, certain pieces of legal arguments are all debated by jurists. It seems that either generations of jurists and others that hoped for help from them were completely wrong, or it is not a prerequisite of legal argumentation that legal statements have a truth value.

In view of the advancement in (philosophical) ethical theory over the past decades, one might be amazed that emotivist ethics, that is, the belief that moral statements express a merely personal attitude (and not a description of this attitude), is still a position frequently adopted in the bioethical literature — for example, so-called situation ethics mainly pursued in health care ethics and feminist ethics.[15]

The result of a moral argument may be called moral knowledge. Based on what has been said on moral statements here, the term *knowledge* would have to be interpreted then in the sense of "*justified* belief" and not in the sense of "*true* justified belief."[16] The dividing line between moral *knowledge* and mere moral *opinion* is therefore not that knowledge is somehow true. I restrict myself to a much less presumptive view: Knowledge, in contrast to opinion, is connected to the pretension that there is some justification procedure that can be used in redeeming the knowledge claim. The aim of this move is to establish a common ground for argument between the partners in conflict. Presupposing stronger conditions for rational statements — like committing oneself to a certain theory of meaning, truth, or theory of science — might hamper this goal.

Ethics and Scientific Method

Since the claim that ethics is of benefit for the mastering of moral problems caused by life-sciences research is often pushed aside by its critics with the claim that ethics is not and cannot be a science, the issue deserves some consideration. One should not, however, invest too much effort in a discussion of whether ethics is a science, since this depends plainly on the definition of *science* — and the whole discussion is in danger of being not much more than a quarrel on words.

In the Anglo-Saxon world the term *science* is normally understood to comprise physics, chemistry, biology, and other disciplines that explore and explain nature. Drawing from the working method of these disciplines, "working scientifically" is understood as the attempt to establish "what is the case" thereby presupposing that a scientific method aims at true, justified beliefs. It is quite obvious that ethics — at least as it is introduced here — cannot be a science in this sense, since it aims at justified beliefs but not true, justified beliefs.

A quick glance at the history of the term *science* shows quite clearly, however, that its meaning has changed considerably over time (Kambartel 1996; Tetens 1999). Dating back to antiquity, the term *scientific* was for a long time interpreted in the sense of "well-founded" or "well reasoned." In this interpretation "ethics" may well be a science. However, this view has been thrust into the background by an empiricist interpretation of science appearing in the nineteenth century and remaining alive through to the current day.

In a time when creationists and other critics deny that the (natural) sciences correctly describe our world and, more importantly, doubt that these disciplines are able to improve our living conditions, it might nonetheless be worthwhile to reassess whether a definition that understands science basically as a rational, methodical procedure is more apt to express the social goals ascribed to the sciences since enlightenment than the empiricist definition of science. Adopting this view would make it more reasonable to "award" ethics the title of a science.[17]

Criteria of Success for a Moral Argumentation

If bioethics is expected to contribute to the solution of social problems caused by the life sciences, proposing a method of bioethics that is only of some theoretical plausibility will not be sufficient. In addition, it has to be shown that bioethics is not just an epiphenomenon of moral debates, so to speak. For the future development of bioethics it will therefore be important to propose criteria that can be used to determine whether a (bio-)ethical theory has been successfully applied to concrete moral problems.[18]

Clarification of Terms. In bioethical debates one is confronted with a large number of terms, such as *human, dignity,* and *nature.* In ordinary communication these terms are assigned a variety of only superficially related meanings, and they have long and disparate histories of discipline-specific use. The explicit agreement on a well-differentiated terminology is, therefore, a prerequisite to securing mutual understanding in bioethical debates. The need for explicitly defined terminology may be briefly illustrated using an example from the debate over research on "human" embryos. The term *human* is sometimes used purely descriptively for the members of the biological species *Homo sapiens,* without any normative conclusions drawn from that. *Human* is also used normatively — that is, with the acknowledgement of an entity as human there are at the same time certain rights awarded. Clearly, depending on the meaning of the term *human,* there will be quite different conclusions developed concerning the moral acceptability of research on human embryos.

Abstracting from this example, a first criterion of success for moral argumentation may be stated:

CS-1   *Clarification of terms:* The success of bioethical argumentation in mastering moral conflicts depends essentially on the nonambiguity and plausibility of the terminology used in the recommendations and their justification.

Rules Governing Moral Argumentation. In addition to developing agreement on the meaning of moral concepts, establishing rules to govern moral argumentation is an important task of bioethics (and of applied ethics in general).[19] Drawing on what Richard Hare (1997, sec. 6.8, 6.9) proposes, the following requirements qualifying rules governing ethical argumentation as correct may be formulated as second criterion of success:

CS-2   *Rules governing moral argumentation:* A successful moral discourse should fulfill the following criteria:

  CS-2.1   *Neutrality:* The ethical theory underlying an ethical argument should be acceptable to all conflict partners.

  CS-2.2   *Practicality:* The result of an ethical argument accepted by all partners in the conflict must be of practical relevance.[20]

CS-2.3 *Incompatibility:* Moral disagreements arising during a moral argument must be disagreements that are not of the "so what" type.[21]

CS-2.4 *Logicality:* The relations between moral statements should be governed by rules of logic, without singling out a special logic as the only adequate one.[22]

Note that bringing substantial presuppositions, that is, morals, into moral arguments likely violates criterion CS-2.1, since one of the discussants might block the argumentation because he or she does not consent to all or some of the presuppositions put forward. This criterion, if adopted, excludes many approaches that are frequently called "ethical theory." John Rawls's "Theory of Justice," for example, probably fails this criterion, since it includes a broad set of "Western" morals in its view on political institutions and their aims. The same may be the case for one of the predominant approaches to bio-"ethics," namely the so-called principlism developed by Tom L. Beauchamp and James F. Childress (2001) and adopted by many scholars. Both approaches are notoriously unsatisfying for those seeking a justification of the morals and/or moral principles used. It becomes quite clear that this is not just an odd complaint of notorious skeptics when one turns to bioethical debates questioning, for example, the principle of autonomy from a non-Western perspective.

Conciliations of Conflicts. It has been said before that mastering moral conflicts is the main aim of bioethics. Based on this a third criterion of success for moral arguments can be phrased:

CS-3 *Conciliation:* A moral argument must serve its goal in society: to reconcile conflicts.

As plausible as this criterion is (if not trivial), it raises some questions: When is a conflict solved? Does the result of mastering a conflict have to be consensus? How is mastering moral conflicts by potentially endless moral argumentation possible in a world with limited time-resources?

Regarding the meaning of *mastering a conflict,* a moral conflict arises when two actors perform actions that are incompatible.[23] This elementary conflict should be differentiated from nonelementary conflicts, where two actors dissent on more than one action and these are incompatible with each other. Though nonelementary conflicts are likely those to be found in real life, they can be reconstructed as several elementary conflicts.

An elementary conflict can be mastered in three ways: First, it can be mastered if the action in question is not performed, in which case the conflict is avoided, not resolved. Second, it can be mastered if at least one actor gives up the aim he or she wanted to realize with a given action. The questions arise, however, whether this would ever happen. Why would an agent give up his or her aim? This scenario can be understood by looking not only at the two incompatible aims but also at the whole set of aims that each actor holds. In this aim structure it is possible to determine higher-order and lower-order aims, where the actor prefers to realize higher-order aims over lower-order ones. From this it follows, finally, that a conflict can be mastered if an agent gives up a lower-order aim in exchange for realizing — as a compensation — another (higher-order) aim now or in the future.[24]

Regarding *consensus in mastering conflicts,* an (elementary) conflict was described as a situation wherein two actors pursue aims that are incompatible with each other. *Consensus* in this view would mean that in the process of mastering the conflict both actors would adopt the same aim. This case will certainly happen in real life, but it seems that another option will be realized too — maybe even more frequently: One or both actors will give up their lower-order aims to retain their higher-order aims. In this case there would be no consensus on those aims that triggered the conflict. It is even not necessary that there is consensus on higher-order aims. The only thing that is required for mastering a conflict is that there does not remain a dissent concerning certain aims (of higher or lower order) that are incompatible with one another. Reaching consensus on aims or norms is therefore neither a necessary nor even a likely precondition of mastering a conflict.

Regarding *ideal discourse and limited time budgets,* so far nothing has been said on the problems of implementing moral argumentation in "real life," that is, under conditions of limited time budgets. In organizing human cooperation there is often simply not enough time to — as it may seem — endlessly discuss the meaning of terms or the likely best way to master a conflict. Consequently, there are a considerable number of cases where, using the ethical method described, conflicts are not or cannot be mastered before a decision needs to be made. It is obvious that in these cases legal discourse will supplement ethical discourse. As a first approximation it may be stated, therefore, that the *differentia specifica* between moral and legal argumentation is that in legal discourse the pragmatic need for shortcuts in decision making on moral issues is taken into consideration by (institutionalized) procedures with a fixed endpoint of discourse.

The Application of Ethical Reasoning in Science and Technology Policy

Having outlined the foundation of bioethics, a description of how moral argumentation may be implemented in the process of political decision-making on science and technology policy is still lacking.

Interdisciplinarity/Ethics as an Auxiliary Discipline

Mastering moral conflicts that are generated by developments in medicine and the life sciences is necessarily an interdisciplinary task — with bioethics playing the part of an auxiliary though necessary science. The (bio-)ethicist is trained in examining the arguments proposed, whether they are valid relative to standards of argumentation or whether they fall back on metaphysical or emotion-based beliefs. Substantial arguments about bioethical problems should be based, however, on knowledge from those sciences in whose domain those problems originated.[25] For example, a discussion of the ethical problems of research on human embryos would be futile without the participation of physicians, jurists, and others, and very likely would not lead to acceptable recommendations for societal regulations. It would be equally futile to assess the moral relevance of, say, foods on nutrition without the methodological support of empirically working social scientists. Finally, it would be futile to develop recommendations for societal regulations, such as the economically efficient and juridically sound allocation of resources in the health care system, without calling on the help of health-economists or jurists.[26]

On the Concept of Bio-Policy

Those who first introduced the term *bio-policy* into the (German) debate used it deprecatingly: Bio-policy is done by those not willing to accept the existing consensus on the principles that should guide moral decision making, it was suggested. (Talking about the "bio-industrial complex" in analogy to the military-industrial complex that is made jointly responsible for World War I is just one example of this offensive style.) But referring to an alleged material value-consensus (be it European, religious, or some other) is faulty for at least two reasons: First, such a consensus does not exist, except in the minds of fundamentalists. This can easily be seen by looking at the different — actually partly contradictory — national regulations on euthanasia and embryonic experimentation. Second, merely claiming that one is in possession of a correct moral is far from arguing for it by using the available method for critically assessing morals: philosophical ethics.

In another reading bio-policy simply designates that part of policy-making that is concerned with questions arising from the life sciences. In this way bio-policy does not differ in any interesting way from monetary or social policy; in view of the importance of the life sciences in our society it seems quite adequate to introduce and use this concept in such a neutral way.

Scientific Policy Consulting

In the case of complex political decisions requiring careful consideration of the pros and cons, politicians usually seek (and find) the advice of relevant experts: tax reform plans no less than climate policies are based at least partially on scientific expertise. Certainly, policy decisions are rarely based on scientific information alone. This does not mean that scientific policy advising is futile; quite to the contrary, it indicates an organizational deficit insofar as it is demanded that policy-making is rational relative to some standards. The idea of supporting policy-making by scientific advisory committees has a very long tradition and was never seriously questioned. *Insofar as bioethics is a rational endeavor, a point can therefore be made in favor of supporting bio-policy-making by ethics committees.* This insight is so intensely disputed due to some serious misunderstandings (and sometimes willful abuse) of institutionalized ethics in policy-making.

As has already been said, scientific advisory committees — whether attached to government bodies or other (professional) organizations — have a long-standing tradition, so arguing in favor of them seems unnecessary. This becomes perfectly clear when one asks whether it would be prudent for Congress, for example, to revise building safety standards spelled out in building regulations without consulting architects and stress analysts. One reason for institutionalizing scientific advice, among others, is pragmatic: on the basis of rules of procedure such committees can produce transparent, long-term reliable advice.

In the last few decades bioethics committees have become abundant at large research institutes and clinics but also as committees attached to government bodies. I have proposed here that the task of ethics is to develop principles for mastering moral conflicts in an argumentative manner. This is basically also the task of institutionalized ethics, that is, ethics committees. Note, however, that there are, at least in Germany, a large number of ethics committees without participating ethicists: though strictly speaking a misnomer, such committees are relatively unproblematic, as long as those committees are intended to enforce the observance of an accepted moral. But with the progressive development of the life sciences those committees are more and more deferred to in cases where traditional morals generate conflicts and need critical evaluation. In these cases ethics committees without ethicists miss their vocation.

Experts, Laypersons, and Lobbyists

Since ethics committees, properly understood, are expert groups, they are frequently blamed for drafting recommendations without a wide enough consultation of the relevant stakeholders (such as patients or the general public). Taking into account the respective opinions of the relevant stakeholders is an important step in preparing a moral recommendation, since it is these mutually incompatible opinions that generate a moral conflict in the first place. But, as said before, the task of ethics (and ethics committees) is to develop recommendations for the solution and avoidance of conflicts, where the aim is recommendations binding for all parties involved in the conflict. This means, however, that the ethical argument should be unbiased and independent of the personal background — including moral opinions (values, preferences) — of those actually performing the assessment. If this can be successfully implemented, ethics committees would not necessarily have to involve all stakeholders; even involving only a very few experts can lead to an ethically justified result. In addition, there is no need, though it is often demanded, for the participation of social groups in ethics committees, whether they be unions, employer's associations, let alone religious groups.

If bioethics is understood as a profession practiced by experts in the same way as the study of radiation effects is a profession practiced by specially trained experts, the institutionalization of bioethics in committees loses its seemingly subversive touch. Ethics committees provide a chance to improve science and technology policy-making by developing reasoned and reliable recommendations concerning the choices and risks of new developments in the life sciences.

The fear that ethics committees might "take over" political power seems to miss the point: An expert committee on, say, radiation protection can assess whether exposure to a certain type and amount of radiation is acceptable relative to some safety standards; in addition, the committee itself can suggest safety standards, but it cannot endow these standards with binding force — except in those cases where the legislators formally delegate some of their regulatory powers to expert committees.

A problem still to be solved in this context is the following: Consulting by expert groups can only be successful when it is possible to differentiate between experts and members of pressure groups (lobbyists). In theory this seems easy: experts are those arguing reasonably consistently within the universally valid standards of the profession they practice. Lobbyists, on the other hand, are those who represent interests on their own or others' behalf, without criticizing, testing, or filtering those interests. Lobbyists may well use professional expertise to reach their goals (for example, opinion polls, marketing tools) and therefore act "rational." But the goals they pursue are not valid universally.

This emphasis on the (theoretical) foundation of validity claims of expert recommendations should not lead to a neglect of the difficulties in the practical implementation of expert procedures: it has become obvious that debates on, for example, changes to international ethical codes like the *Declaration of Helsinki* involve highly political conflicts of interest, where the involved parties persist in their claims, whether they are justified or not (Schüklenk and Ashcroft 2000). To overcome these practical problems we likely need further institutional arrangements, such as that an expert committee must work independently of interested professional organizations.

In conclusion one can say that ethics consulting by experts neither endangers the representation of the citizen nor necessarily reduces transparency. The claim that ethics committees lack democratic legitimacy is weak. In contrast, "moral consulting" is afflicted by a defect. Moral decisions cannot be made on behalf of the individual by an expert. Therefore, what lacks democratic legitimacy is moral consulting by lobbyists. The representation of the citizen would be weakened if decisions were made not by the political functionary elected by the citizenry but by a nonelected committee making final decisions on moral issues. But even if a moral committee does not decide such matters, what competence in moral issues does an employer or unionist have, that makes him or her better qualified than an elected official?

NOTES

I would like to express my gratitude to Georg Kamp for valuable help in clarifying my thoughts while preparing this chapter. I also thank Mauro Dorato for his sharp yet always constructive criticism during the Pittsburgh-Konstanz colloquium in October 2002.

1. See also Morscher, Neumaier, and Simons 1998, xiv–xviii.
2. There is considerable dispute inside theology on the relationship between moral theology and moral philosophy and on the role of the revelation for determining what one ought to do; see, for example, Vorgrimmler (2000, 431–32).
3. Part of the material in this section was developed in Gethmann and Thiele (2003).
4. I hope that the differentiation between *moral* and *morals* does not give too much trouble to those who are native speakers of English. It appears that in English there is only the plural *morals*, whereas in German there is the singular *Moral* and the plural *Moralen*, as is the case in Latin, with *mos* and *mores*. To use the term *morality* in this context does not seem to be adequate since it connotes that there are morals, not the actual set of moral convictions a certain individual cherishes—which I call a "moral." See also Frankena (1973, 4–9).
5. What it means for a recommendation to be *satisfying* is discussed later.
6. For further reading (though not necessarily in line with the position I develop in this chapter), see Gillon (1998), Korff (1998), Kuhse and Singer (1999), Beauchamp and Walters (1999), Engelhardt (2000), Düwell and Steigleder (2003).
7. Martin Carrier suspected that this way of describing ethics might make it difficult to demarcate ethics from "couples therapy." Though I think that ethics is mostly engaged in mastering conflicts that arise in a public sphere and couples therapy is not, I do not deny that there may be certain similarities in the aims that therapists and ethicists pursue: both seek to make people communicate successfully (again). However, a therapist working toward this goal will rarely use argumentation theory and other philosophical methods, training his or her psychological capacities instead.
8. Certainly there is the difficult case of "moral fanatics" who would rather give up their cooperative aims than their moral principles (see Hare 1981, chap. 10).
9. This formulation is very close to Lorenzen and Schwemmer (1975, 150).
10. See also Gethmann (1982, 1992).
11. Lorenzen and Schwemmer (1975, 150) argue, however, that such a view on ethics results in a plain restriction of the tasks traditional ethics was pursuing.
12. This classification is adopted from Hare (for example, Hare 1997, pt. 2).

The latter position has been specially worked out as "value-ethics" by Max Scheler and Nicolay Hartmann (Schnädelbach 1999, chap. 6). Another intuitionist ethicist is G. E. Moore.

13. See Kamp (2001, 23–28). In a historical study Rainer Hegselmann (2002, 7–46) has argued that intuitionist ethics has been the main opponent in contrast to which the logical empiricists have developed their view on ethics — a view that is likely responsible for the aversion with which ethics is still viewed by many philosophers. It should be added that in recent times some scholars once more give credit to the view that there are moral facts; see, for example, McDowell (2001).

14. Hare (1952) has probably been the first scholar that argued in this way. For a recent in-depth elaboration of this issue see Kamp (2001); For my position it is not necessary to presuppose that moral statements cannot have a truth value. It is sufficient to claim that moral statements do not necessarily have to have a truth value to be part of a reasoned moral argument. With this concession I hope that those who claim that moral statements do have a truth value as well as those who claim otherwise may nonetheless adopt the method of ethics I propose here.

15. See, for example, Kuhse (1998).

16. In this context there is often talk of "values" thought to be a component of moral knowledge. Following the theory of value ethics, there is a tendency to maintain that claims of values have a truth value. The term *value* is a favorite word in bioethical debates, but since the term is understood frequently in the way just sketched, I will not use it here.

17. When I presented my intuitions on this issue during the Pittsburgh-Konstanz colloquium, Mauro Dorato suspected that I was using the term *science* "just as an unnecessary honorific title, used to advertise bioethics as a 'respectful' field that would benefit science policy and the public." Though I agree that my point is not easily defended, I think it is of the same kind and importance as the refutation of the claim that physics, chemistry, and biology are not any more valid than the belief in witchcraft.

18. The following owes much to Hare (1997) and Kamp (2001). An alternative approach has been sketched by Morscher Neumaier, and Simons (1998).

19. For a reconstruction of moral discourse in terms of theory of action, see Gethmann (1982, 1992)

20. Descriptive types of ethics fail this requirement.

21. As would be the case with some types of subjective naturalism.

22. Some forms of emotivism might not fulfil this requirement. *Logicality* should be understood in a weak sense, insofar as this requirement does not presuppose that moral argumentation must necessarily be governed by the rules of any specific type of philosophical logic — be it classic, intuitive, and so on. Kamp (2001, 210–14) suggested a "maxim of use-orientation" (*Maxime der Usus-Orientiertheit*): The relations between moral statements should be governed by inferential rules as they are used in our common language.

23. In addition, a conflict may arise when two actors pursue norms (justifying aims) that are not compatible with one another. Since what is said about conflicts of aims applies analogously for the collision of norms, I will concentrate on the former.

24. See Lorenzen and Schwemmer (1975, 166–68); Schwemmer (1976, 33, 34). It is not claimed, however, that establishing a hierarchy of aims will always be sufficient to master a conflict.

25. At what point bioethicists have to draw on the expertise of other non-philosophical disciplines for their considerations depends on the competence of the bioethicists—more specifically on their expertise *beyond* the method of applied philosophy.

26. For further reading, see Grunwald (1999) and Decker (2001).

REFERENCES

Beauchamp, T. L., and J. F. Childress. 2001. *Principles of biomedical ethics*. New York: Oxford University Press.

Beauchamp, T. L., and L. Walters, eds. 1999. *Contemporary issues in bioethics*. Belmont, CA: Wadsworth.

Decker, M., ed. 2001. *Interdisciplinarity in technology assessment: Implementation and its chances and limits*. Berlin: Springer.

Düwell, M., and K. Steigleder. 2003. *Bioethik. Eine Einführung*. Frankfurt: Suhrkamp.

Engelhardt, H. A. T. 2000. *The philosophy of medicine and bioethics: an introduction to the framing of the field*. Dordrecht: Kluwer, 1–15.

Frankena, W. K. 1973. *Ethics*. Englewood Cliffs, NJ: Prentice Hall.

Gethmann, C. F. 1982. Proto-Ethik. Zur formalen Pragmatik von Rechtfertigungsdiskursen. In *Bedürfnisse, Werte und Normen im Wandel*, ed. H. Stachowiak and T. Ellwein, vol. 1, *Grundlagen, Modelle, und Prospektiven*, 113–43. München: Fink/Schöningh. (English: 1989. Protoethics: Towards a formal pragmatics of justificatory discourse. In *Constructivism and science: Essays in recent German philosophy*, ed. R. E. Butts and J. R. Brown, 191–220. Dordrecht: Kluwer.)

———. 1992. Universelle praktische Geltungsansprüche. Zur philosophischen Bedeutung der kulturellen Genese moralischer Überzeugungen. In *Entwicklungen der methodischen Philosophie*, ed. P. Janich, 148–75. Frankfurt: Suhrkamp.

Gethmann, C. F., and F. Thiele. 2003. Grundlagen der ethischen Bewertung der Gentechnik. In *Handbuch der Molekularen Medizin: Molekulare und Zellbiologische Grundlagen*, vol. 1, ed. D. Ganten and K. Ruckpaul, 711–34. Heidelberg: Springer.

Gillon, R. 1998. Bioethics overview. In *Encyclopaedia of applied ethics*, vol. 1, ed. R. Chadwick, 305–17. San Diego, CA: Academic Press.

Grunwald, A., ed. 1998. *Rationale Technikfolgenbeurteilung. Konzepte und methodische Grundlagen*. Berlin: Springer.

Hare, R. M. 1952. *The language of morals*. Oxford: Clarendon Press.

———. 1981. *Moral thinking: Its levels, method, and point*. Oxford: Clarendon.

———. 1997. *Sorting out ethics*. Oxford: Clarendon Press.

Hegselmann, R, ed. 2002. *Moritz Schlick: Fragen der Ethik*. Frankfurt: Suhrkamp.

Kambartel, F. 1996. Wissenschaft. In *Enzyklopädie Philosophie und Wissenschaftstheorie*, vol. 4, ed J. Mittelstraß, 719–21. Stuttgart: Metzler.
Kamp, G. 2001. *Logik und Deontik*. Paderborn: Mentis.
Korff, W. 1998. Einführung in das Projekt Bioethik. In *Lexikon der Bioethik*, ed. W. Korff, L. Beck, and P. Mikat, 7–16. Gütersloh: Gütersloher Verlagshaus.
Kuhse, H. 1998. A nursing ethics of care? Why caring is not enough. In *Applied ethics in a troubled world*, ed. E. Morscher, O. Neumaier, and P. Simons, 127–42. Dordrecht: Kluwer.
Kuhse, H., and P. Singer, eds. 1999. *Bioethics: An anthology*. Oxford: Blackwell Publishers.
Lorenzen, P., and O. Schwemmer. 1975. *Konstruktive Logik, Ethik, und Wissenschaftstheorie*. Mannheim: Bibliographisches Institut.
McDowell, J. 2001. *Mind, value, and reality*. Cambridge, MA: Harvard University Press.
Morscher, E. 2000. *Angewandte Ethik*. Salzburg: Forschungsberichte und Mitteilungen, Forschungsinstitut für Angewandte Ethik, Universität Salzburg.
Morscher, E., O. Neumaier, and P. Simons. 1998. Introduction. In *Applied ethics in a troubled world*, ed. E. Morscher, O. Neumaier, and P. Simons. Dordrecht: Kluwer.
Schnädelbach, H. 1999. *Philosophie Deutschland 1831 bis 1933*. Frankfurt: Suhrkamp Verlag. (English: 1984. *Philosophy in Germany 1831–1933*. Cambridge: Cambridge University Press.)
Schüklenk, U., and R. Ashcroft. 2000. International research ethics. *Bioethics* 14 (2):158–72.
Tetens, H. 1999. Wissenschaft. In *Enzyklopädie Philosophie*, ed. H. J. Sandkühler, 1763–73. Hamburg: Meiner.
Vorgrimmler, H. 2002. *Neue theologisches Wörterbuch*. Freiburg: Herder.

14

Knowledge and Control:

On the Bearing of Epistemic Values in Applied Science

Martin Carrier
Bielefeld University

The Primacy of Applied Science

Among the general public, the esteem for science does not primarily arise from the fact that science endeavors to capture the structure of the universe or the principles that govern the tiniest parts of matter. Rather, public esteem — and public funding — is for the greater part based on the assumption that science has a positive impact on the economy and contributes to securing or creating jobs. Consequently, applied science, not pure research, receives the lion's share of attention and support. It is not knowledge that is highly evaluated in the first place but control of natural phenomena. The relationship between science and technology is widely represented by the so-called *cascade model*. This model conceives of technological progress as growing out of knowledge gained in basic research. Technology arises from the application of the outcome of epistemically driven research to practical problems. The applied scientist proceeds like an engineer, employing the toolkit of established principles and bringing general theories to bear on technological challenges. The cascade model holds that promoting epistemic science is the best way to stimulate technological advancement.

The preference granted to applied science increasingly directs university research toward practical goals; not infrequently, it is sponsored by industry. Public and private institutions increasingly pursue applied projects; the scientific work done at a university institute and a company laboratory tend to become indistinguishable. This convergence is emphasized by strong institutional links. Universities found companies in order to market products based on their research. Companies buy themselves into universities or initiate large-scale contracts concerning joint projects. The interest in application shapes large areas of present-day science.

This primacy of application puts science under pressure to quickly supply solutions to practical problems. Science is the first institution called on if advice in practical matters is needed. This applies across the board to economic challenges (such as measures apt to stimulate the economy), environmental problems (such as global climate change or ozone layer depletion), or biological risks (such as AIDS or bovine spongiform encephalopathy, BSE). The reputation of science depends on whether it reliably delivers on such issues. The question naturally arises, then, whether this pressure toward quick, tangible, and useful results is likely to alter the shape of scientific research and to compromise the epistemic values that used to characterize it.

There are reasons for concern. Given the intertwining of science and technology, it is plausible to assume that the dominance of technological interests affects science as a whole. The high esteem for marketable goods could shape pure research in that only certain problem areas are addressed and that proposed solutions are judged exclusively by their technological suitability. That is, the dominant technological interests might narrow the agenda of research and encourage sloppy quality judgments. The question is what the search for control of natural phenomena does to science and whether it interferes with the search for knowledge.

On the Relation between Knowledge and Power

Underlying these considerations is the notion that pure and applied science differ in nature. Otherwise, the endeavor to clarify the relationship between the two would not make sense. In contrast to this presupposition, it is argued in some quarters that science is intrinsically practical. The only appropriate yardstick of scientific achievement is usefulness or public benefit. In this vein, Philip Kitcher denounces the view that the chief aim of science is to seek the truth as the "myth of purity" and advances the contrasting idea of a "well-ordered science" whose sole commitment is satisfaction of the preferences of the citizens in a society (Kitcher 2001, 85–86, 117–18). "Well-ordered science" is an ideal Kitcher wants scientists to pursue; it is not intended as description of reality. Still, his approach squares well with a widely shared feeling that practical use or technology is what science is essentially all about. Given a commitment of this sort, no significant distinction between theory and practice or between knowledge and power can be drawn.

It is true, indeed, that claims to the effect that the touchstone of epistemic significance is practical success originate with the Scientific Revolution. However, it is also true that these commitments largely remained mere declarations. Take Christopher Wren who was familiar with the newly discovered Newtonian mechanics when he constructed St. Paul's Cathedral. The Newtonian laws were deemed to disclose the blueprint of the universe, but they were unsuitable for solving practically important problems of mechanics. Wren had to resort to medieval craft rules instead. Likewise, the steam engine was developed in an endless series of trials and errors without assistance from scientific theory (Hacking 1983, 162–63). Thermodynamics was only brought to bear on the machine decades after its invention was completed (see below). This gap between science and technology is not completely filled today. Theoretical work on cosmic inflation will hardly ever bear technological fruit. Such work is driven exclusively by curiosity; pure knowledge gain is the focus. Conversely, screening procedures in the development of medical drugs possess neither theoretical basis nor theoretical import. In such procedures, cellular or physiological effects of substances are detected and identified by using routine methods. They involve a more sophisticated form of trial and error.

I conclude that there have been and still are purely epistemic and purely practical research projects. Neither is science inherently practical, nor is technology inherently scientific. This means that the distinction between basic research and technology development needs to be upheld. And this, in turn, suggests that the relationship between seeking the truth and developing some useful device merits a more thorough consideration.

The connection between science and technology becomes manifest only in the nineteenth century. The now familiar pattern that a technological innovation emerges from the application of scientific theory is an achievement that succeeds the Scientific Revolution by roughly two centuries. The cascade model is intended to capture this more recent relationship between scientific knowledge and practical use. The idea is that technological progress grows out of scientific theorizing. Technology really is applied science. This model can be taken to involve the twofold claim of substantive and causal dependence of technology on science; that is, the operation of some technical device can be accounted for within a relevant theory, and the device was developed by applying the theory. According to the cascade model, the logical and the temporal relations run parallel: theoretical principles are formulated first; technical devices are constructed afterward by spelling out consequences of these principles.

Overtaxing Science by Application

Scientists themselves are sometimes found articulating concerns to the effect that applied science involves a reduction of the high methodological standards that characterize basic research. Organizations of scientists—like the German *Stifterverband für die deutsche Wissenschaft*—warn that science is too heavily put into the service of social, political, and economic interests. This outside pressure is assumed to weaken the epistemic control procedures inherent in respectable research. Insufficiently confirmed ideas are aired publicly and passed prematurely as scientific knowledge. Others agree with the diagnosis but find nothing wrong with the methodological changes possibly induced by application orientation. It is argued that the need to address complex issues tends to increase the uncertainty of scientific claims, encourages interdisciplinary cooperation, and promotes inclusion of social values into research. But these features are not considered a cause of concern; they are instead welcomed as signaling a many-voiced, less monolithic, and more democratic structure of science (Funtowicz and Ravetz 1994).

Let me address the issue of possible methodological differences between applied and basic science more systematically. A plausible mechanism of methodological erosion appears to be operative in applied science. The pressure on scientists to quickly deliver useful recipes drives science toward addressing increasingly complex issues. The reason is that phenomena and effects that can be put to technological use are only rarely easy to capture theoretically. Applied science is denied the privilege of basic research to select its problems according to their tractability. Applied science cannot confine itself to areas in which effects appear without distortions, idealizations hold, and approximations work satisfactorily. Instead, applied science has to face intricacy and lack of perspicuity (Krohn and van den Daele 1997, 194–95, 199–200).

The need to address complex situations tends to overburden science. It is plausible that scientists respond to such excess demands by adopting *tentative epistemic strategies*. This methodological erosion is likely to extend to *theory structures* and *criteria of judgment*. Application-dominated science could consist of diverse collections of clusters, each of which might comprise specific assumptions unconnected by a unifying theoretical bond; that is, models invoked in applied science could be at once internally heterogeneous and differ substantially from one another. Likewise, scientists might appraise suggested models exclusively on their potential for intervention. It would be sufficient for accepting a proposed model that it enables control. The goal to reach an understanding of natural phenomena would be abandoned. Elucidating causal mechanisms or embedding them in the system of knowledge might be taken as an epistemic luxury that applied research cannot afford.

There are indications for such an overtaxing of science by its application and the ensuing resort to tentative epistemic strategies. Three methodological features can be observed whose combined or marked appearance tends to be characteristic of applied science: *local models* rather than unified theories, *contextualized causal relations* rather than causal mechanisms, *real-world experiments* rather than laboratory experiments conducted for answering theoretical questions.

First, at least some areas of applied research contain collections of specific models of narrow scope, only loosely tied together by shared principles. This is true, for instance, of large parts of cognitive psychology or economics. The number of model-specific assumptions is large, and it is these particular claims that bear the explanatory burden. Second, not only do such models lack common principles, but the causal claims entertained therein are often highly contextualized. Contextualized causal relations only hold under typical or normal conditions and leave the pertinent causal processes out of consideration (see below). A third characteristic of applied science is the high portion of "real-world experiments" that investigate the entities or devices in question during their practical use. Such experimental strategies are the result of limitations in accounting for the pertinent issues within a more comprehensive theoretical framework. The complexity of the situation renders laboratory studies useless. Theoretical understanding is not sufficiently advanced to allow the transfer of insights gained under controlled conditions to the situation at hand. Thus, real-world experiments are employed for gaining usable knowledge in the first place, not for testing hypotheses suggested by a theoretical approach. The eventual predominance of such Baconian experimental strategies with their emphasis on exploration and heuristics reveals that science operates at its epistemic limit. It tends to be overburdened by the control requirements set on science by the economy or politics.

The ecological management of waste deposits is a case in point. The large number of pertinent chemical reactions and their interaction with particular soil structures make it impossible to anticipate the environmental impact of certain substances by drawing on laboratory studies. Rather, it is only the operation of a waste deposit under real-life conditions that allows one to estimate those parameters that are essential for the appropriate construction and safe operation of the site (Krohn 1997, 76; Krohn and van den Daele 1997, 203–6). Another example is software development. No single person and no small group of persons is able to understand the totality of processes that occur in a complex computer program. This is why changes and adaptations in one part of the program are frequently followed by a completely unforeseen malfunction in a different part. The early market release of such programs is, at bottom, a real-world experiment: programs are intended to mature with the customer. That is, mistakes are identified through customer complaints and fixed in a piecemeal fashion. Wide areas of application-dominated research are characterized by such strategies of trial and error.

Prospects of a Self-Contained Applied Science

Such methodological features of applied science nourish the suspicion that science suffers from the grip of practical demands. Theories of fundamental science are widely expected to excel in virtues like explanatory power, predictive force, or unifying capacity. By contrast, applied science seems to be characterized by local models, contextualized causal relations, and exploratory experimentation. The explanatory and unifying bearing of theory apparently fails to extend to technological challenges. On the face of it, applied science is methodologically deficient as compared to fundamental science.

On the other hand, the cascade model of applied science provides a safeguard against worries of methodological decline. The cascade model assumes a unidirectional dependence of applied science on fundamental research. If this were appropriate, pure science would not be at risk by the emphasis on application. First, on this model, pure research is the presupposition for enduring technological innovation so that a unidirectional dependence of technology on science ensues. Second, fundamental knowledge cannot be produced according to a preconceived plan; it needs creativity or the spark of genius. Moreover, the relevance of a piece of fundamental knowledge for a given practical challenge cannot reliably be estimated in advance. It follows that fundamental research has to be conducted in a broad fashion and without bearing a particular application in mind. Conversely, if you narrowly focus on technological development, you dry up the potential for future technological progress. The concentration on practical problems would eventually deprive science of its capacity to solve practical problems. Given the unidirectional dependence of applied science on fundamental research, the search for control demands the pursuit of knowledge. Pure science would be safe. As a consequence, the demanding methodological standards, supposed to be inherent in basic research, would be maintained in applied science.

But philosophers of science and scientists alike have cast doubt on the cascade model. Nancy Cartwright is a prominent example. In her view, overarching laws or high-brow theories do not gain access to concrete phenomena. The laws of physics fail to account for large parts of the physical world. She approvingly refers to an example of Neurath's who pointed out that Newtonian mechanics is embarrassingly silent on the path of a thousand-dollar bill swept away by the wind in Vienna's St. Stephen's square (Cartwright 1996, 318). In her conception of a dappled reality, the only way to take a grip on the phenomena is by making use of local models that are tightly locked onto a particular problem. Descriptive adequacy only extends to small-scale accounts; comprehensive theories inevitably lose touch with the phenomena. The patchwork quilt, not the pyramid, symbolizes the structure of scientific knowledge (322–23).

The physicist Silvan Schweber advocates a similar ontological picture—if for different reasons. Schweber attributes a layered structure to reality, with each such layer arising from the complex organization of matter. A relation of "objective emergence" is thought to obtain among the pertinent properties, according to which the study of each level of complexity is as fundamental in its nature as any other. In spite of their hierarchical order, the higher-level account is not just the applied science of one level below. That is, the theories that address the different levels of organization are essentially decoupled from one another. Elementary-particle physics has virtually no implications for atomic physics or solid-state physics (Schweber 1993, 35–36, 38).

The relationship between fundamental theory and practical application, as envisaged by these positions, is at variance with the cascade model. The message is that pure science largely fails to meet applied challenges. Practical problems are to be attacked directly and without a detour through fundamental research. There is only a minor knowledge transfer from the constituents of matter to complex aggregates. Fundamental truths do not produce technological spin-offs.

The adoption of this point of view brings a mixed message in its train. It entails, on the one hand, that the structure of pure science is much more akin to the assumed nature of applied science than anticipated. Part of the worries mentioned before draw on the alleged contrast between the overarching and unified theoretical edifice of pure science, and the collection of heterogeneous small-scale models supposedly distinctive of applied science. However, on the sketched point of view, pure science may likewise comprise a collection of scattered and divergent approaches.

Actually, an attenuated version of Cartwright's position can pass as the now standard view of theory structure in epistemic science. This standard view grants that general theories capture concrete phenomena, to be sure, but only if assisted by auxiliaries peculiar to the situation in question. Models are mediators between overarching theories and concrete phenomena, and these models involve divergent assumptions. At most, these models stand in a relation of family resemblance (as the semantic view has it); but they may even comprise contradictory principles and fail to form a coherent whole (Hacking 1983, 216–19; Morrison 1998, 70–81).

Abandoning the cascade model is good news, therefore, in that this move mitigates the perceived contrast between pure and applied science and makes applied science look more like respectable epistemic research. Even granted that applied science only contained problem-specific approaches that employed heterogeneous assumptions, applied science would not necessarily lag behind epistemic research in methodological respects. The bad news is that the safety cord is thereby lost. I explained that within the cascade model the best way to promote applied science is to foster pure science. If the cascade model is relinquished, applied science can take care of itself. Given the predominance of practical concerns, it becomes a live option that pure science is pushed toward the sidelines.

Applied Science and Contextualized Causal Relations

These considerations reveal that the relationship between epistemic and applied science is a matter of debate. The best way to resolve such a contentious issue is to address concrete cases. Let me examine the example of *contextualized causal relations*. I mentioned before that the restriction to such relations is one of the possibly characteristic methodological features of applied science. In reality, causal relations typically form complex networks. A cause does not produce an effect completely on its own but through a sequence of intermediate events. Causality operates through processes or mechanisms. Moreover, a multiplicity of causes may bring about the same effect. The *epistemic attitude* attempts to understand this web of parallel series of concatenated events. The *pragmatic attitude* aims at intervention and control. Pragmatic investigations focus on cause-effect relationships that prevail under "typical" conditions and are thus usually sufficient for bringing about the effect. Such contextualized causal relations incorporate a large number of presuppositions; they are thus confined to "normal" circumstances. In other words, such relationships are massively hedged with *ceteris paribus* clauses.

Starting a car is an example. A pertinent causal relation is *turning the ignition key sets the motor in motion*. But this effect only follows if a chain of intermediate processes manifest itself. The causal capacity of the ignition key considered in itself fails to bring about the desired effect. The turning of the key closes an electric circuit, which instigates a series of other events that eventually produce the start of the engine. The given causal relationship passes in silence over these further conditions; they are tacitly incorporated. The act of turning is neither sufficient nor necessary for starting a car. It does not produce the effect but merely triggers it.

The turning of the ignition key and the car's starting are connected by a contextualized causal relationship. Such a relationship only holds if the embedding causal network operates as tacitly presumed. Contextualized causal relationships are confined to normal circumstances and consequently suffer from exceptions. They collapse if one of these tacit clauses is no longer fulfilled. If the spark plug is wet, all turning serves for nothing. In addition, such relationships fail to elucidate the connection between cause and effect. They leave the underlying causal chains unaddressed.

The concentration on contextualized causal relationships is sometimes considered a hallmark of the "practical sciences." Take the philosopher of history Robin Collingwood who already in 1940 characterized this pragmatic attitude by its renunciation of a comprehensive analysis of the pertinent causal network: "If I find that I can get a result by certain means, I may be sure that I should not be getting it unless a great many conditions were fulfilled; but so long as I get it I do not mind what these conditions are. If owing to a change in one of them I fail to get it, I still do not want to know what they all are; I only want to know what the one is that has changed" (Collingwood, quoted in Fox Keller 2000, 142). This means that the scope of the practical sciences is narrow. Knowledge is important only to the extent that it is necessary for enabling successful intervention. The underlying assumption is that it is possible to intervene dependably on the basis of fragmentary knowledge. Technology is taken not to rely essentially on a deeper understanding of the pertinent generalizations.

Biotechnology and the Life Sciences

In the course of the last two decades, the life sciences have developed into one of the most important fields of applied science. Biological processes have been subjected to a large number of human interventions. Molecular genetics or genetic engineering play a leading role in this area. Its pivot is the assumption of a fixed connection between genes and organismic features or between genes and proteins. On the basis of such unambiguous connections, genetic engineering nourishes the hope for — or the fear of — specific interventions in organismic processes.

Genetic engineering frequently draws on contextualized causal relationships. The way the so-called eyeless gene is put to use is a case in point. This gene is found in the species *Drosophila*, the common fruit fly, but it possesses homologs in mice and men. It controls for the morphogenesis of the eyes of flies. If the operation of the gene is blocked or lost, no eyes are formed — which is why the gene is somewhat misleadingly called "eyeless." If the homologous mouse gene is implanted and expressed in *Drosophila*, it instigates the formation of fly eyes, not mouse eyes.

The expression of the eyeless gene in suitable tissue is sufficient for eye formation; that is, eyes can be generated by appropriate stimulation in the legs or wings of flies. This is the reason for calling eyeless the "master control gene for eye morphogenesis." But the eyeless gene only sets off a complex series of intertwined genetic processes that only in their entirety control eye formation. This is evidenced by the mentioned fact that the homologous mouse gene stimulates the expression of fly eyes in fly tissue. The eyeless gene operates as a trigger that needs the appropriate causal environment in order to become effective (Fox Keller 2000, 96–97). In its causal role, this gene is comparable to the ignition key. When an ignition lock is removed from a fashionable convertible and built into a battered pick-up truck, the turn of the key sets the truck engine in motion, not the motor of the convertible.

The upshot is that the identification of eyeless allows the control of eye morphogenesis without theoretical understanding of the underlying processes. Results of that sort are taken as a basis for a declaration of independence of technology from science that is enunciated within the biotechnologist camp. The background to this judgment of irrelevance is the widespread assessment that the life sciences are presently undergoing a revolution regarding the understanding of the interrelationship between genetic and somatic processes: genomics is being replaced by proteomics. Genomics was governed by the idea of genetic determinism according to which the properties of a cell are fixed by its genes. This is contested by the rival approach of proteomics, which emphasizes the fact that many cell properties are the result of an intricate interaction among proteins. To be sure, proteins are produced by genes, but they are still subject to generalizations of their own; generalizations, that is, that are not encoded in DNA. In principle, limitations of the scope of the genome arise from the fact that external factors may influence the expression of genes in a cell. Such external factors escape the grip of the genome of the corresponding cell, but they contribute to determining which genes are active and which remain switched off. This means that a full account of cell properties cannot be reduced to the DNA-level but has to address, at least in part, protein interaction directly. This transition from genomics to proteomics is viewed by the biologist Richard Strohman as a "coming Kuhnian revolution" (Strohman 1997, 194). Similarly, Evelyn Fox Keller considers the twentieth century as "the century of the gene." After the turn of the century we realized the deep rift between genetic information and biological function (Fox Keller 2000, 8).

The crucial aspect is that this thesis of the conceptual and theoretical insufficiency of genomics frequently goes along with holding fast to the notion of the gene for pragmatic reasons. In spite of her judgment that this notion is fundamentally flawed, Fox Keller continues to accept *gene* as a useful concept that refers to "handles" for intervention. Genetic manipulation is suitable for producing effects in a uniform, predictable fashion, albeit the reliability of the relevant causal relationships is constrained to particular sets of conditions (Fox Keller 2000, 141–42). The same separation between scientific adequacy and technological importance is advanced by William Bains, who works as a consultant for biotechnology companies. Bains argues that genetic determinism is superseded and discarded in bioscience but rules unquestioned in biotechnology. It is true, the assumption of a close connection between gene and organismic property is scientifically dubious, but it still constitutes a lever for opening the black box of life. Genes and their products are tools for achieving technological progress and for bringing about intended effects. Scientific truths are not necessary for this purpose. This is shown, among other things, by the fact that wrong conceptions as to the mechanisms of sulfonamids, aspirin, or penicillin prevailed for a long time. Technology aims at practical success, which is achieved by the identification of levers to pull and switches to press. Focusing on truths about the underlying causal processes is bound to generate confusion in biotechnology (Bains 1997).

The common ground of all these judgments is that biotechnology rests on contextualized causal relationships whose appropriateness is independent, in large measure, from the truth of more fundamental, high-brow theories. In particular, successful intervention need not rely on disclosing the relevant causal mechanisms. Knowledge and control are taken to be decoupled, and the commitment to truth is sacrificed for the capacity of intervention.

Bioscience as a Basis of Biotechnology

It follows that there are indications of the attitude of the "practical sciences." That is, some projects in the realm of "research and development" are confined to the mere production of an effect. If the intervention is successful, no further questions are posed. Restriction to and satisfaction with contextualized causal relationships bear witness to the suspected methodological decline: no elucidation of the causal mechanism is attempted, no integration of the local relationship into the wider body of knowledge is sought. Taken together with the prima facie evidence for the other methodological features presented earlier, namely, local modeling and real-world experimentation, there is reason to believe that tentative epistemic strategies are employed in applied science.

The follow-up question is whether attitudes of this sort prevail in applied science. A definite answer to this question requires a more extensive survey of applied research projects (I am involved in one[1]). But a preliminary answer can be given by drawing on circumstantial evidence and an in-principle argument. I restrict consideration to the role of contextualized causal relations in biotechnology and suggest that both sources render it unlikely that appeal to such relationships dominates applied science.

Relevant circumstantial evidence is provided by the fact that we do witness the mentioned revolution in bioscience, namely, the replacement of genomics by proteomics. If it is true that biotechnology is fine with genomics, and if it is further true that biotechnology sets the research agenda, no such revolution should occur. As long as genomics successfully underwrites biotechnological practice, and we are told it still does, genomics should be here to stay. The fact that a conceptual transition is nevertheless underway testifies to the fact that there is more to applied science than contextualized causal relationships.

A large number of observations point in the same direction. Research projects in biotechnology frequently aim to identify the network of generalizations and causal mechanisms that bear on a particular causal relationship. For instance, the development of new medical drugs often draws on advanced knowledge of the underlying cellular processes. Drugs are designed by relying on insights about the molecular nature of the disease. In this area progress in therapy is sought by way of clarifying the molecular mechanisms whose malfunction is the physiological basis of the disease.

Let us look a little more closely at the example of antibiotics research. This endeavor is certainly technological in nature; the driving force is the development of effective treatment. It is true that the first step into this field was taken without any deeper knowledge. Alexander Fleming discovered the antibiotic efficacy of penicillin by chance. Substances with similar capacities can be found by screening, that is, again, without understanding the underlying processes. However, the next step requires appeal to more fundamental knowledge. Namely, sustained treatment with antibiotics produces resistance. The efficacy of antibiotics decreases since bacterial variants are selected whose molecular structure is less vulnerable to antibiotic action. For instance, some antibiotics interfere with the formation of the bacterial cell wall. Resistant bacteria employ peptides of a slightly changed structure for building up their cell walls so that the molecules of the antibiotic no longer combine with sufficient intensity to these peptides. Disclosing the details of this mechanism makes it possible to develop countermeasures. The molecules of the antibiotic are modified such that they connect tightly to the changed wall peptides and thus regain their original efficacy.

I take this pattern to be typical. Contextualized causal relations, like the antibiotic efficacy of penicillin, may indeed provide a basis for some limited intervention in an organismic process. But if distortions arise, such contextualized relations with their long list of hedging clauses attached are of no use anymore. Additional influences invalidate one of these tacit clauses so that the generalization fails to become manifest. We are faced with an exception to the generalization. Reinstating its validity or regaining the power of intervention requires getting such disturbing influences under control. This can be achieved either by checking these influences or by blocking their adverse effects. In most cases, the latter route is the only way to go, and proceeding on this path demands taking account of the relevant mechanism. Consequently, maintaining the efficacy of a procedure in the presence of distortions requires theoretical understanding of the causal chains at hand. Sophisticated intervention requires deeper insights. At the advanced technological level presently addressed, there is no way to decouple control from knowledge.

Application Innovation or How to Hook Up Technology with Science

The import of this conclusion is not confined to antibiotics research nor to biotechnology. Rather, applied science in general is bound for methodological reasons to transcend itself and to grow into fundamental science. This is the in-principle argument mentioned earlier. Lack of a deeper understanding eventually darkens the technological prospects. But theoretically understood causal relations provide many more opportunities for intervention and control than do contextualized causal relationships. The theoretical explanation or integration heightens the chances to bring other factors to bear on the process at hand and to twist the latter so that it delivers more efficiently or reliably what is demanded. A good theory is extremely practical.

On the one hand, applied science is tempted by its practical nature to be satisfied with contextualized causal relations. The challenge is to figure out which switches are to be pressed to achieve some desired effect. Theoretical understanding is not among the objectives. On the other hand, applied science does not yield—at least not unreservedly or not always—to the temptation to merely give generalizations heavily laced with *ceteris paribus* clauses. There is a subtle mechanism at work that keeps basic research right within the applied ballpark. Practical challenges often bring fundamental problems in their wake. Such challenges cannot appropriately be met without treating these fundamental problems as well. Basic research is among the results of applied science; the former is engendered by the latter. A consequence is that innovative explanatory approaches that are relevant for basic research frequently arise within applied research projects. This feature I call *application innovation*. It involves the emergence of theoretically significant novelties within the framework of use-oriented research projects.

A large number of such use-oriented projects in the life sciences address questions of theoretical impact. They furnish innovative solutions to problems of epistemic relevance and are thus different from sciences like astrophysics or paleontology, on the one hand, which are pursued out of pure curiosity, as well as from engineering or the development of marketable goods, on the other. The inquiry into the genetic and enzymatic processes that control cell division is an example of research at this intermediate level. The knowledge is about the foundations of cellular reduplication, but at the same time it offers prospects of intervention if cell division gets out of hand. These studies were conducted in order to achieve a more effective cancer therapy (which practical interest is underscored by the fact that the Nobel prize for medicine was awarded to the corresponding scientists). Yet they also generated fundamental insights into the mechanism of cell division.

Likewise, the revolutionary conception of prions was elaborated in the practical context of identifying infectious agents. Prions were conceived of as infectious proteins that assumedly reproduce without assistance of nucleic acids (DNA or RNA). The initial aim of the investigation was to gain useful knowledge about the sheep disease scrapie; later the bovine ailment BSE moved toward center stage. But the impact of this study was a deep-reaching transformation of biological concepts. The pursuit of a practical question produced a profound theoretical innovation regarding biological reproduction.

This is by no means an exclusive feature of present-day science. Rather, since the nineteenth century theoretical knowledge proves increasingly helpful for tackling practical problems. One of the early exemplars of application innovation is Sadi Carnot's 1820 treatise on the motive power of heat. Its declared aim was to analyze and improve the workings of the steam engine. To accomplish this aim, Carnot introduced the seminal concepts of thermodynamic cycle and thermal efficiency. These concepts are of enduring theoretical significance.

Cases of this sort do not instantiate the familiar pattern that technology draws on scientific knowledge. To be sure, this cascade pattern is manifest in many cases. But my point is that the converse is also true and that knowledge of nature's workings may grow out of the attempt to master technological challenges. Application innovation is seen at work in such cases.

Eventually, the very formation of science as a rule-governed epistemic enterprise goes back to this pattern of problem generation that directs scientists from application to foundation. Answers to practical problems were sought by elaborating theoretical accounts. A purely theoretical interest was only rarely dominant in the history of science. What mattered was knowledge useful for the betterment of the human condition. The scientific method with its stringent demands on the acceptance of hypotheses contributes to ascertaining the reliability of practically relevant knowledge. For instance, unified explanation and causal analysis rank highly among the widely shared methodological virtues. These virtues codify in the first place what knowledge or understanding is all about. We understand a phenomenon when we are able to embed it in a nomological framework, and we grasp a causal relationship when we can account for the process leading from the cause to the effect.

These same virtues are also essential for successful research on practical matters. I mentioned that the theoretical integration of a generalization or the clarification of intermediate processes facilitate intervention. Furthermore, restricting contextualized causal relationships to a narrow class of conditions seriously impedes their transfer to other situations. By contrast, theoretically understood relationships can more easily be generalized and applied to a wider range of conditions. Consequently, the latter sort of relationship is more useful in practical respects.

It follows that the primacy of application need not pose a threat to the epistemic dignity of science. Theoretical unification and causal analysis are inherent in both pure and applied science, because such methodological virtues promote understanding and intervention at the same time. Applied questions are treated best when they are not treated exclusively as applied questions. Thus, scientific method provides a safety cord that secures the epistemic respectability of applied science.

Application innovation thus entails a partial vindication of the cascade model. While Cartwright is right in her claim that concrete problems need to be treated using local means, such narrow approaches naturally grow into more comprehensive ones and stimulate the treatment of fundamental issues. This second aspect of a natural tendency toward enlarging the explanatory scope matches well with the cascade model in one respect but fails to fit it in another respect. Scope enlargement agrees with the cascade model in that practical answers are best given by exploring theoretical problems. Technical devices are often developed by elaborating consequences of some theory. But this same aspect is at odds with the cascade model in that the theory need not precede the application. The cascade model says that the most effective way to foster applied science is to support pure science. My argument is intended to show that stimulation may proceed in the opposite direction. Temporally speaking, insights of epistemic relevance may grow out of practical projects. In contrast to the cascade model, the logical and temporal relationships between science and technology sometimes run in opposite directions.

Conclusion

The message here is twofold. While it is true, on the one hand, that applied science is more prone to methodological sloppiness than epistemic research, one the other hand, there are shared values that act as a safeguard against wholesale methodological decline. It is to be granted that the dominance of science by practical purposes brings adverse effects in its wake. The agenda of science is largely monopolized by technological challenges, and the importance of research done in private companies compromises the public accessibility of knowledge. As to the first, application dominance markedly influences the selection of issues treated by science. Regarding the second, the somewhat excessive secrecy that covers a great deal of industrial research tends to reinstate the traditional separation between scholarship and society at large that was gradually resolved by the Scientific Revolution. Not all worries about an application-dominated science can be dispelled.

However, there is a more positive message as well: application dominance eventually poses no serious threat to the epistemic respectability of science. Applied science is no less concerned with trustworthiness and reliability than pure science, and methodological values like theoretical unification and causal analysis are tried and tested means for accomplishing these practical purposes as well. Whereas applied science brings to bear additional practical values like efficiency, low cost, or environmental friendliness, there is continuity among the epistemic values accepted in both fields. In the end, the striving for control is not likely to override the commitment to knowledge.

NOTES

1. This research is part of the project "Toward the Knowledge Society" pursued at the Institute for Science and Technology Studies at Bielefeld University and sponsored by the Volkswagen Foundation and the Deutsche Forschungsgemeinschaft.

REFERENCES

Bains, W. 1997. Should we hire an epistemologist? *Nature Biotechnology* 15:396.
Cartwright, N. 1996. Fundamentalism versus the patchwork of laws. In *The philosophy of science*, ed. D. Papineau, 314–26. Oxford: Oxford University Press.

Fox Keller, E. 2000. *The century of the gene*. Cambridge, MA: Harvard University Press.
Funtowicz, S. O., and J. R. Ravetz. 1994. Uncertainty, complexity, and postnormal science. *Experimental Toxicology and Chemistry* 13:1881–85.
Hacking, I. 1983. *Representing and intervening: Introductory topics in the philosophy of natural science*. Cambridge: Cambridge University Press.
Kitcher, P. 2001. *Science, truth, democracy*. Oxford: Oxford University Press.
Krohn, W. 1997. Rekursive Lernprozesse: Experimentelle Praktiken in der Gesellschaft. *Technik und Gesellschaft* 9:65–89.
Krohn, W., and W. van den Daele. 1997. Science as an agent of change: Finalization and experimental implementation. *Social Science Information* 36:191–222.
Morrison, M. 1998. Modelling nature: Between physics and the physical world. *Philosophia Naturalis* 35:65–85, ed. B. Falkenburg and W. Muschik.
Schweber, S. S. 1993. Physics, community, and the crisis in physical theory. *Physics Today*, November, 34–40.
Strohman, R. C. 1997. The coming Kuhnian revolution in biology. *Nature Biotechnology* 15:194–200.

15

Law and Science

Eric Hilgendorf
Department of Law, University of Würzburg

The relationship between law and science can be investigated from various perspectives (Mnookin 2002). One could inquire into the ways the law restricts and promotes science or the degree to which modern science has altered the law. Biotechnology, for example, has challenged the adequacy of existing legal structures. Another route of inquiry is how science has affected the laws of evidence and the limits of judicial competence. One could follow the history of the connection between law and science or the varied interpretations of these concepts within each country's legal culture. In Germany and many other European countries, the study of law *(Rechtswissenschaft)* is commonly regarded as a science (Curran 2002, 3); the relationship between law and science is not more problematic than the relationship between, say, sociology and science or biology and science. In comparison, no one in the United States or Britain would call a university-educated lawyer a "scientist."

Naturalism in Law and Philosophy

Although all of these routes of inquiry are legitimate, I have opted to approach the topic from a different, more "philosophical" vantage point by asking what the relationship is between law and its systematic study, on the one hand, and law and philosophical naturalism on the other. The general naturalistic program draws from the interpretation of contemporary empirical sciences, in particular, from the natural sciences and medicine. It is hardly an exaggeration to say that naturalism is the world view of modern science. Moreover, it is closely connected with the idea of a "unity of sciences," a concept that goes back to the Vienna (and Berlin) Circle of Logical Empiricism. The term *Wissenschaftliche Weltauffassung*, which Rudolf Carnap, Hans Hahn, and Otto Neurath used in 1929 to characterize the intellectual basis of logical empiricism (Schleichert 1975, 201–22), is closely related to today's concept of philosophical naturalism.

In German, and, indeed, European, legal philosophy, the term *naturalism* is employed in a variety of cases. In older German legal and jurisprudential literature, naturalism was viewed very negatively, often in the context of positivism, mechanism, and even nihilism. Unfortunately, most of these terms were not precisely defined. Naturalism was equated with an alleged "hostility to morality" and warnings of its morally corrosive effects were issued. Reference was made to "the desert-like character of a theory which possesses none of the moral starting points fundamentally necessary for human society" and the tendency of naturalism to "leave both people and their world completely deserted, internally and externally" (von Hippel 1967, 253). It comes as no great surprise that naturalism was taken to be partially responsible for not only the defeat of the Weimar Republic but also the reign of violence witnessed during the National Socialist period.

Since the 1970s, the usage of such moralistic slogans has died down. In most modern theoretical and jurisprudential works, the term *naturalism* does not appear at all, though occasionally the idea of naturalism can still be found in academic legal discussions. In such instances, naturalism, having been coupled with "blind" causality and causal theory, is typically being contrasted with "teleological" or "value-related" jurisprudence. We can conclude that the term as it is portrayed in older German and indeed European literature, has a pronounced polemic character.

In modern philosophy, naturalism stands for the idea that, fundamentally, all phenomena have a natural explanation (Nagel 1956). Viewed from this angle, naturalism appears to be somewhat of a simple reformulation of Ockham's Razor as a program of science. Recently, a German philosopher of science put forward the following minimalistic program of naturalism: "Only so much metaphysics as necessary.... Minimal realism, so that it is possible for a world to exist without human beings. . . . Primacy of inanimate matter/energy. . . . Construction of real systems from real components. . . . No instances that transcend experience. . . . No miracles. . . . Even the intellectual achievements of man do not extend beyond nature" (Vollmer 1995, 40).

The modern conception of philosophical naturalism has not yet been integrated into European jurisprudence nor, as far as I am able to judge, into British/U.S. jurisprudence. Some legal theorists' reservations could stem from the fact that some of the points just mentioned — minimal realism, for example — are completely self-evident to lawyers. Other postulates, in comparison, do not have any relevance to lawyers' work, for example, the axiom of inanimate matter/energy. Legal philosophers and lawyers could also be reluctant to openly accept naturalism because of the older perception of naturalism as a morally and politically dangerous theory.

Lawyers and legal philosophers often simply repeat arguments already submitted by philosophers fifteen or twenty years earlier. Why exactly this happens is in itself an interesting problem. Many jurists have an inclination to isolate themselves from other disciplines. Interdisciplinary work may also be lacking because of widespread confusion over philosophical terminology, especially when different philosophical schools are involved. Some schools, for example, German "discourse ethics" or French "postmodernism," even seem to regard confused language as a sign of intellectual superiority.[1] Moreover, there is a tendency in contemporary philosophy to dissociate from material issues. Philosophers prefer instead to respond to philosophers' assessments or to philosophers' critiques of other interested parties' assessments. It is not so much the "strife of systems" (Rescher 1985) but this common modern trend of philosophers formulating meta-meta-metatheories that very much hinders interdisciplinary work.

Historical Links Between Law and Science

As I see it, law and science are more closely linked than they first appear. Increasingly, legal ways of thinking, both in Europe and the United States, are stimulated by scientific developments and often change course as a result. This was already apparent at the turn of the twentieth century. A distinguishing feature of this period is the movement toward empirical questions. These questions materialized in such influential trends as the jurisprudence of interests (Rudolf von Jhering, Philipp Heck), ethnological legal research (Albert H. Post), legal sociology (Max Weber, Arthur Nussbaum), criminology, and research regarding the purposes of punishment (Franz von Liszt). Criminal law in particular is indebted to naturalism for the strong humanizing impulses it brought. These tendencies entered the mainstream of American jurisprudence initially through the writings of Oliver Wendell Holmes, Karl L. Llewellyn, and Roscoe Pound. Scholars emigrating from Nazi Germany (such as Ernst Rabel and Max Rheinstein) brought their ideas on methodology to America, where they blended with traditional American legal thought.

Medicine is a perfect demonstration of how the employment of a truly naturalistic methodology fuses with a human-oriented discipline. Modern medicine's immense successes since the nineteenth century have only been possible thanks to both naturalism and the decisive application of scientific methods. It goes without saying that we expect good doctors not only to try to cure us in the best way technically possible but also to possess understanding and a basic human touch. However, this is no argument against naturalism: On the contrary, a doctor who wants to help his or her patient will use the best, that is, the most efficient, methods available, the methods approved by science. Medicine clearly shows how a naturalistic method can be combined with the humanistic way of thinking.[2]

I seek in this chapter to demonstrate that a naturalistic program can be integrated into jurisprudence and without any significant changes to the current legal order. I begin by discussing questions of extra-legal values and norms and their relevance to the legal system. I will then look at two rather different concepts, both of which are very important to most legal systems: human dignity and causality. I discuss their compatibility with the naturalistic program. I conclude with a look at the techniques of legal argumentation and, more specifically, of statutory interpretation.

Values and Norms

Questions of Value in Law

Questions concerning the foundation of values and norms are actually more common in jurisprudence than one would expect, judging from a superficial look at legal practice. The most important example is the creation of new legal norms. It is obvious, however, that the moral contents of legal norms differ immensely. Minimal changes to the highway code or the law of land registry, for example, have much less of a morality aspect than, say, consumer protection cases or criminal law on abortions. In the realm of international law, the International Criminal Court statute was drafted in the summer of 1998. In doing this it was necessary to formulate criminal law norms that could be binding worldwide. It goes without saying that technical legal problems and problems concerning fundamental norms and values came up and had to be dealt with in the process.

Not only do questions concerning values arise in the creation of new legal norms; they also sometimes arise when attempting to abandon existing norms. Such problems became important in Germany after the downfall of the Nazi and Communist regimes. In both systems preexisting legal norms seemed to sanction conduct that was not simply illegal but also scandalously immoral from the vantage point of the new legal framework. The German courts treat such norms as nonlegal and therefore invalid. In doing so, the courts draw reference to the legal philosopher Gustav Radbruch, who shortly after World War II argued that some Nazi laws were so immoral that they lacked legal character (1946, 105–8). The main problem with this statement is the question of how *extreme morality* should be defined without simply deferring to the moral prejudices of a given culture or specific groups.

A third example of the importance of fundamental values and norms in jurisprudence is the interpretation of general phrases like "*in good faith* and *public policy.* The law seems to refer to a positive standard but does not reveal where to find it. The problem associated with the interpretation of such general clauses is a particularly clear example of how the application of written law is hardly ever truly a reconstruction of the standard being set by fundamental values and norms. Hans Kelsen has already acknowledged that the application of law nearly always has a political element as well (Kelsen 1960, 346–54).

In a nutshell, questions of value and moral norms are nearly ubiquitous in law. How are we to find standards of value and norm that are more than just the result of cultural or group prejudices?

Sociobiological Reasoning in Law?

Rather than presenting an overview of twenty-five hundred years of moral philosophy, I want to show what a naturalistic answer to questions of values and norms in law could look like. Often those making naturalistic moral assertions claim that these are supported by sociobiology. But just because a natural behavior schedule is *factual*, it does not follow that we *should* act in accordance with it. The naivety with which some sociobiologists repeatedly confuse facts and norms is astonishing.

The case of sociobiology, in my opinion, demonstrates further problems associated with naturalistic thought patterns. It appears to me that in some cases the recourse to "natural" biological factors has not been well thought out. In many cases, sociobiologists merely assert "reproductive interests." Let us take a modern textbook example of sociobiology from Franz Wuketits (1997, 174). Here the author attempts to reconcile the decision of a man to live celibate as a priest with the assumption of overwhelming and ubiquitous reproductive interest. The sociobiologist writes that celibate priests make a choice to live without reproduction and families of their own. However, the priestly life is taken to have certain advantages. Indeed, a priest's reputation benefits not only himself but also his parents, brothers, and sisters. Therefore, the sociobiologist argues, the family of the priest profits from his celibacy. This is clearly saying that the priest's reputation is promoting the overall reproductive chances of his siblings. The priest thereby still manages to help pass on those genes that he has in common with his siblings.

To my mind, the sociobiological argument is not very convincing. What would happen if potential partners of a priest's sister were actually deterred by the sheer fact that her brother was a priest? The only path remaining open to such a sister would lead to the convent. This would appear to be somewhat of a detriment to her reproductive success.

Naturalistic thought does not, however, necessarily lead to reductionism and oversimplification. Explanations of physical phenomena, such as personal preferences and personal values, may take biological factors into account without neglecting such other factors as religion and social influences. The decision taken by a person to become a priest is definitely influenced by many factors. This decision, like all other human decisions, can be analyzed naturalistically.

There are other areas in which biological research into behavior could influence the law. For one thing, biological research could make searching for intercultural or even global values easier. The validity of such values could be laid out in a general agreement of international consent. Moreover, behavioral biology could uncover genuine intercultural methods of law reinforcement and could devise symbols that support the acceptance of legal norms in a universally understandable manner. Biologists exploring human dispositions could explore ways of altering those that contravene legal norms.

It is clear that the law must also pay attention to human nature: law should not demand the impossible (*ultra posse nemo obligatur*). Research that explores the factual possibilities of realizing legal norms also falls within the scope of behavioral biology. It seems possible for the biology of human behavior to trace hitherto undoubted norms back to their biological roots, thereby making possible a critical analysis. One last point: That human behavior is based on biological dispositions and subject to the process of natural evolution by no means excludes the fact that, nowadays, people can actually decide against their own biological disposition and, over time, change their own natural development.

An Interest-Oriented Concept of Value Foundation

There is another, indirect way of making sociobiological findings and empirical anthropology fertile ground for lawyers: A legal norm is substantiated by the proof that the norm is a sufficient, possibly even a necessary condition for the creation or protection of a positively valued situation. The question of which conditions are positively valued obviously depends on the wishes and interests of the deciding individuals. A value is arrived at by abstraction from positively valued situations or "states of the world." It follows then that the term *value* as used here is subjective or relative (that is, relative to people and their needs).[3]

Moral judgments are not generally arrived at completely freely and arbitrarily but are traceable to a scale of valuation. Valuations of individuals are therefore assumed to be more or less consistent. The valuation scales in question often vary by only a few degrees from individual to individual. Even within a large group, uniform values are often to be found. From this we can conclude that, even in large groups or whole cultures, and possibly stretching as far as worldwide, clear homogenous valuations, and therewith values, can be pinned down. Certain conditions are valued universally, such as self-survival and personal health, as well as the lives and health of close relatives. Also valued is the satisfaction of core physical requirements, like nourishment and rest. Furthermore, a minimum level of material security would also seem to be universally valued.

Since humans are a product of evolution, there is an obvious tendency to try to explain our needs and interests biologically. As far as life, health, and core physical requirements this does not seem problematic. But this means that certain value dispositions, and consequently values, are the result of natural evolution. Research into and systematization of such natural interests and value standards falls under the umbrella of biology and empirical anthropology. In this sense, there is no categorical or logical gap between the factual and the normative.

It is difficult to avoid using the words *natural law* when talking about this subject. Still, there are significant differences between the outline that I have given of universally accepted norms and values and what we understand under the classical models of "natural law": Biology does not prescribe individual norms, but outlines natural interests, which can help to formulate a common, possibly globally acceptable canon of values. The implementation of such values into moral or legal norms affords the legislator a high level of freedom. Moreover, such norms cannot simply be arrived at by following natural guidelines; an actual *decision* must be made regarding the realization of a "naturally" existing disposition. It appears that not all natural interests are worth realizing. Moreover, it is obvious that throughout all cultures religion and morality mold people's conduct when they are assigning value. The positive valuation enjoyed by certain behaviors, therefore, consists by no means solely of "natural interests."

In spite of this, recourse to natural needs and interests is a means the naturalistic legal philosopher could choose to use in solving law's normative problems. We can ask whether legal and moral norms are compatible with our own natural tendencies and, furthermore, consider the extent to which they provide a suitable channel for conveying our natural interests. It seems, therefore, that you can be a naturalist and still a legal philosopher with more to offer than pure logical formalisms and simple value skepticism.

Naturalism in Legal Dogma

In order to review the possibility of transferring naturalistic thinking into law, it is not sufficient to simply consider questions of values and norms, for this is a matter of philosophy of law rather than of law itself. In this section, I discuss the possibility of interpreting practical legal concepts in a way compatible with the naturalistic program.

Human Dignity

Human dignity (*Menschenwürde*) plays an immense role not only in legal academia but also in daily newspapers and the philosophical literature. Human dignity, in spite of or perhaps owing to its widespread status, is extraordinarily ill-defined and metaphysically charged. For this reason, the concept is an ideal candidate for clarification on naturalistic lines.

The principle of human dignity provides a basis for state constitutions, including the German constitution (Art. 1, para. I, *Grundgesetz*), but it is also often involved in much-debated legal-political issues. Misuses of the human dignity concept are common. Since the widely held understanding of human dignity is so broad and overinterpreted, it is, for example, applied to debates over genetic cell research, where even double or quadruple human cells are viewed as having human dignity. Therefore, all research into such cells is categorically ruled out. On the other hand, such research seeks to find cures that will help improve the human dignity of sick and disabled people. It is obvious then that the argument of "human dignity" can be used both for and against genetic cell research, which boils down to the fact that the argument is hardly of any worth in this debate. To take the idea of "human dignity" seriously, we must make the concept more precise.

The interpretation of human dignity in European jurisprudence has two roots, the first being the theological tradition of "man" as "image of God"; the second being the idea of individual autonomy developed during the Enlightenment. According to the definition of human dignity given by German courts, human dignity is infringed if "a person is significantly disparaged as an object or a mere means." The Kantian source of this definition is unmistakable. In many more popular statements, an infringement of human dignity is defined as the "instrumentalization" of an individual, that is, as the using of him or her solely as a means to achieve a certain purpose.

However, the ban placed on instrumentalization is much too narrow to exhaust all cases that breach human dignity. Let us take the example of a fanatical dictator who tortures to death the last remaining member of a resistance organization in the cellar of his palace. In this case, the dictator is not seeking to misuse his victim for any specific purpose. The breach of human dignity witnessed here does not arise because the event has a purpose-means structure. Human dignity is infringed because of the *effects* that such treatment has on the victim, who has been locked up, stripped of all rights, and tortured. The expression "instrumentalization of others" is therefore inapplicable as a general criterion or definition of the violation of human dignity.

From a naturalistic standpoint, human dignity can be defined by an ensemble of subjective rights. These are oriented toward basic human requirements and interests. One could call this the "ensemble theory" of human dignity (Hilgendorf 1999b, 148–50). The requirement for the minimum goods necessary to exist (food, air, space) corresponds to "the right to a minimum material existence." The interest in fundamental freedom possibilities corresponds to "the right to autonomous self-development," and the interest in freedom from pain corresponds to "the right to be without pain." The interest in keeping personal information confidential corresponds to "the right to a private sphere," and the interest in intellectual liberty (freedom from brainwashing, for example) corresponds to "the right to intellectual-psychological integrity." The interest in a safe, fundamental legal status corresponds to "the right to fundamental legal equality," and the interest in not being publicly humiliated corresponds to "the right to minimal attention."

The introduction of these seven subjective rights as a definition of human dignity can be interpreted as a *naturalization* of the idea of human dignity. It is, perhaps, worth casting a glance over the advantages and disadvantages of this conception: On the pro side, there appears the clear advantage of making a widely held concept more precise. Whether there has been a violation of human dignity can be confirmed in a comparatively stringent way by going through the list of subjective rights. But the contra side of the argument also requires attention: The precision gained in terminology corresponds, of course, to a loss in ways to interpret human dignity. In this way, the concept loses part of its emotional power. This is the price we have to pay if we want human dignity to become a truly legal, not just philosophical, concept.

Even more problematic is the question of which type of requirements are formulated to underpin and explicate the content of human dignity: Are we matter-of-factly talking about natural requirements and how such requirements can be used for a "naturalization" of the term *human dignity*? Or are the needs mentioned above culturally superformed or indeed completely dependent on culture and therefore fundamentally culture-specific? For example, as far as the interest in protecting the private sphere is concerned, we are indeed talking of a culturally superformed need that has only become apparent in recent years, primarily in our Western industrial and media-oriented societies. Other needs, such as the interest in being free from severe pain, seem to have stronger biological roots. As a whole, it seems clear that the naturalization of human dignity brings advantages as well as considerable problems.

Causality

A further test for the naturalistic program in jurisprudence is the concept of causality. The naturalistic position has been characterized by the thesis that all phenomena can be causally explained, and thus the principle of causality cannot be broken by a supernatural power. Empirical science has the task of researching natural causal links. In law, causality problems play an immense role when the causal relationship between an action and its result must be examined. In addition to carrying out the offence, the perpetrator has to actually achieve a factual result (such as to cause the *death* of a person or to *damage* something).

In German law, the test for determining whether a causal relationship exists is called the *conditio sine qua non* formula (comparable to the British/American *but for* test; see Robinson 1997, 153–54). An action is considered causal when the existence of the end result depends on the action's having been performed. In other words, a causal relationship is said to exist between A and B, if A is a condition necessary for the fulfillment of B.

This model of causality is compatible with the naturalistic position (Hilgendorf 1999a). Supernatural instances do not play any role in the legal examination of causality. Even causal relationships between psychic and physical factors (and vice versa) are considered unproblematic. The philosophical problems of mental causality have not yet reached European jurisprudence. Nevertheless, the question of whether A was a necessary condition for B cannot be answered without considering their empirical connection. In simple cases, judges are helped here by their personal life experiences: If a ball hits a window, and the glass in the window subsequently breaks, then the causal relationship is identifiable without scientific examination.

It is a different matter if the issue at stake is, say, whether a particular medication has physically harmed its consumer. In such cases, medical experts must deliver their opinions in order to rule out other possible causes that may come into consideration. In the Contergan case, for example, which appeared before the German courts in the early 1970s, it was argued that physical deformities suffered in connection with the drug Thalidomide could possibly be the result not of Thalidomide but of the atom bomb tests carried out during the 1950s and 60s. In that case, the consumption of Thalidomide would not have been a necessary condition for the physical deformities. Such problems nevertheless do not hinder the naturalistic interpretation of the concept of causality. Thus, the legal interpretation of causality is perfectly compatible with the naturalistic way of thinking.

Legal Method and Naturalism

Finally, the methods of legal argumentation in German and European law crucial to a lawyer's everyday work need to be investigated vis-à-vis the question of naturalism.[4]

Statutory Interpretation

The four classical legal methods of statutory interpretation, otherwise known as "the interpretation canons," are the grammatical, the systematical, the historical, and the teleological methods. Grammatical methodology looks at the choice of words and syntax of the written legal norm in question to determine how it should be interpreted. This essentially philological process is in itself compatible with the concept of naturalism. The is true as well for the systematical method of interpretation. Here, the aim is to apply the general sense of the term enunciated in other legal texts to the interpretation of the terminus in question. Historical interpretation methods look at the views historically held by the legislature: What was the problem at hand? Which solution did the legislature intend? Such questions are empirical and therefore generally create no great problems in terms of naturalistic methods.

More difficult is judging the teleological method. Many textbooks on legal methods claim that teleological methods are used to research the objective sense of the law. This explains why this method is often referred to as an "objective teleological" method of interpretation. But what exactly this means is not clear. On the one hand, it can refer to the sense of a law or legal phrase being a clear statement of the aims of the legislature initiating it, in which case, the objective teleological method of interpretation coincides with the historical method. On the other hand, it is possible that the teleological method is borne purely by the purpose of those applying the law. It is their own purpose that they are finding in the law. A law text in itself has independently no "sense." To speak of "objective sense" brings us back to the mysticism of Hegel's terms, to which the objective teleological method of interpretation is in fact closely related.

On closer inspection of the practical use of teleological arguments, the confusion seems to dissolve. In fact, the debate almost always centers on practical considerations that argue both for and against possible variations in linguistic interpretation. Such variations are then normally supported by considerations of the empirical consequences of a certain interpretation of the law. The choice to interpret words in a given way mostly comes down to prognoses. Understood in this sense, teleological methods of interpretation also conform with naturalistic methodology.

Further methods used in statutory law and its application normally fall within the categories outlined by the four canons. Application to higher-level legal systems, such as European law or international law, requires conceptual and logical consistency. The consideration of how problems are solved in other legal systems, so-called comparative law, can essentially be carried out with empirical methods, even if questions of interpretation do often crop up. Legal sociology, legal anthropology, and legal history are all empirical disciplines.

The Application of Law

My final question is whether the practical application of law itself is compatible with a naturalistic program. To apply law practically is not a purely logical process, meaning that statues alone do not drive the decisions of those who are applying the law. More often than not, the law suggests a mere framework within which those entrusted with applying the law are free to fill out. In doing so, all kinds of considerations are logically possible and actually realized. This means, of course, that concepts contrary to naturalism can play a role as well.

In German law, no explicit norm compels those applying the law to conform to the standards of naturalistic argumentation. But it is unimaginable that someone applying the law today would explicitly refer to the will of a god, goddess, or other supernatural entity.[5] A judge arguing openly in a religious or some other metaphysical way would be regarded as irrational. The wide interpretation of fundamental legal concepts permits that such considerations, if they in fact were to arise, could simply be hidden behind such terms as *"human dignity."* As for legal practice, however, I am not aware of any contemporary judgment in Germany that oversteps the bounds of naturalistic method. In other words, one could say that the application of law may be, but is not necessarily, compatible with the naturalistic program.

Conclusion

My aim in this chapter has been to argue that modern law is compatible with the naturalistic program of science

Here I summarize my main points. Empirical biological research into human value dispositions, that is, our natural tendencies to value things or situations as "good" or "bad," can help us to find a basis for transcultural legal values. Therefore, there is no need to turn to non- or antinaturalistic methods, which are notoriously problematic. It is possible to reinterpret such vague concepts as *"human dignity"* in ways that are compatible with the naturalistic program. *Causality* as the term is already used by lawyers accords with the naturalistic program. The is true as well for the methods of legal argumentation, at least as far as the methods of statutory interpretation are concerned (the problem of case law argumentation was left undiscussed; presumably, the result would not be different). Lawyers applying the law can (and sometimes will) hold views that are not compatible with the naturalistic program. But in modern legal practice, non-naturalistic beliefs are more hidden than explicit.

NOTES

1. All the more important is the adherence to scientific standards in philosophy; see Wolters (1994).
2. The combination of naturalism and humanism is also to be found in the Vienna (and Berlin) Circle of Logical Empiricism; see Hilgendorf (1998).
3. For a detailed discussion of value theory, see Rescher (1969).

4. For the British/U.S. perspective on legal argumentation, see Charter and Burke (2002).

5. After World War II, German courts cited Roman-Catholic Thomist philosophy of natural law in prosecuting Nazi crimes. The judicature was supported by prominent legal philosophers such as Helmut Coing and Heinrich Rommen (a good example is Rommen 1964).

REFERENCES

Carter, L. H., and T. F. Burke. 2002. *Reason in law.* 6th ed. New York: Addison Wesley Longman.

Curran, V. G. 2002. *Comparative law: An introduction.* Durham, NC: Carolina Academic Press.

Hilgendorf, E. 1999a. Causality in penal law: Explanation or understanding? A sketch. In *Actions, norms, values: Discussions with Georg Henrik von Wright,* ed. G. Meggle, 265–72. New York: Walter de Gruyter.

———. 1999b. Die missrauchte Menschenwürde: Probleme des Menschenwürdetopos am Beispiel der bioethischen Diskussion. In *Annual review of law and ethics,* vol. 7, ed. B. S. Byrd, J. Hruschka, and J. C. Joerden, 137–58. Berlin: Duncker & Humblot.

Hilgendorf, E., ed., 1998. *Wissenschaftlicher Humanismus. Texte zur Moral- und Rechtsphilosophie des Wiener Kreises,* vol. 12 of *Haufe Schriftenreihe zur rechtswissenschaftlichen Grundlagenforschung.* Freiburg: Haufe.

Hippel, E. von. 1967. Zur Überwindung des Naturalismus in Recht und Politik. In *Festschrift für Fritz von Hippel zum 70. Geburtstag,* ed. J. Esser and H. Thieme, 245–62. Tübingen: J. C. B. Mohr (Paul Siebeck).

Kelsen, H. 1960. *Reine Rechtslehre. Mit einem Anhang: Das Problem der Gerechtigkeit.* 2nd ed. Wien: Franz Deuticke.

Mnookin, J. 2002. Science and law. In *The Oxford companion to American law,* ed. K. L. Hall, 714–18. Oxford: Oxford University Press.

Nagel, E. 1956. Naturalism reconsidered. In *Logic without metaphysics and other essays in the philosophy of science,* ed. E. Nagel, 3–18. Glencoe, IL: Free Press.

Radbruch, G. 1946. Gesetzliches Unrecht und übergesetzliches Recht. *Süddeutsche Juristenzeitung* 1:105–8.

Rescher, N. 1969. *Introduction to value theory.* Englewood Cliffs, NJ: Prentice Hall.

———. 1985. *The strife of systems: An essay on the grounds and implications of philosophical diversity.* Pittsburgh, PA: University of Pittsburgh Press.

Robinson, P. H. 1997. *Criminal law.* New York: Aspen.

Rommen, H. A. 1964. In defense of natural law. In *Law and philosophy: A symposium,* ed. S. Hook, 105–21. New York: New York University Press.

Schleichert, H., ed. 1975. *Logischer Empirismus: Der Wiener Kreis. Ausgewählte Texte mit einer Einleitung.* München: Wilhelm Funk.

Vollmer, G. 1995. Was ist Naturalismus? Eine Begriffsverschärfung in zwölf

Thesen. In *Auf der Suche nach der Ordnung. Beiträge zu einem naturalistischen Welt- und Menschenbild*, ed. G. Vollmer, 21–42. Stuttgart: Wissenschaftliche Verlagsgesellschaft.

Wolters, G. 1994. Scientific philosophy: The other side. In *Logic, language, and the structure of scientific theories*, ed. W. Salmon and G. Wolters, 3–19. Pittsburgh, PA: University of Pittsburgh Press.

Wuketits, F. M. 1997. *Soziobiologie. Die Macht der Gene und die Evolution sozialen Verhaltens*. Heidelberg: Spektrum.

Index

accuracy, as value of science, 58, 63, 66, 76n4, 127
actions, 109n14; on insufficient evidence, 73–74, 167n9; prudential *vs.* evidential motivations for, 60, 74; and theories, 46, 253; values and, 28, 162–63, 165; *vs.* beliefs, 250–51. *See also* applications
adaptationism, 192–94; classes of activity in, 207, 214n17; controversy over, 205–6; Evita model for, 194–201, 213nn6, 11, 214n12; legitimacy of, 202–5; measurement of, 206–8
agroecology, 38, 49nn14–15
antiballistic missile (ABM) program, 228, 246–48
application innovation, 289, 291
applications, of knowledge, 84–85, 98–103, 119
applications, of research, 38, 41, 105, 119; in biotechnology, 284–86; contextualized causal relationships in, 283–84; esteem for, 275–76; focus on, 34, 291; influence on methodology, 278–81; legitimation of, 41, 44, 46; models of, 279–80; risks of, 118, 120–21; social values and, 44, 46; theories' relation to, 281–82; values in, 1, 40, 52, 70–75; *vs.* pure research, 277–78. *See also* technology
Atomic Energy Commission, General Advisory Committee of, 224–26
"Attacks on Science: The Risks to Evidence-based Policy" (Rosenstock and Lee), 236

autonomy, 48n7; in government-science relations, 222, 239; of methodology, 24, 43, 46; as value of science, 24–25

Bains, William, 286
Bazelon, David, 230
beliefs, 262; effects of, 60, 65–66, 73–74, 181–82; objectivity of, 176–82; self-confirming, 181–82, 184
Bellarmino, Roberto, 64
Bethe, Hans, 225
bioethics, 256; advising government policies, 268–69; moral argumentation in, 264–65; objectives of, 257, 259–60, 263, 265–67; plurality of theories in, 259–60; sources of conflict in, 261–62; in stem cell research, 115–16, 302
biology, 192–94, 205, 214n18
biotechnology, 284–88
Boltzmann, Ludwig, 68–69
Boyd–Putnam rule, 17–18
"brain death," 5–6
Braithwaite, John, 91
Brooks, Harvey, 230
Brown, C. Titus, 213n6
Bush, Vannevar, 222–24, 240n1

capitalism, global, 25
Carnap, Rudolf, 8, 15
Cartwright, Nancy, 281–82, 291
cascade model, of technology and science, 275–78, 281–83, 290–91

311

causal relations, contextualized, 279–80, 283–86, 289, 291
causality, 56–57, 290–91; and nonepistemic values, 73, 248–50; in philosophy of law, 304–5
Church, the. *See* religion
classification/categorization, 5–6
cognitive, the, 53; relation to the social, 22, 85, 87, 129–30. *See also* values
cognitive/epistemic values, 19–21, 52; cognitive but nonepistemic, 19; definition of, 47n3, 53; dominance of, 39–41; and empirically equivalent theories, 59–60; examples of, 32–33, 183; and inductive risk, 249–50; manifestation of, 31–32; nonepistemic and, 52–53, 60, 74–75; and objectivity, 7–8; in scientific practices, 4, 30–32, 43, 48n6, 80; social dimensions of, 79, 134; in theory acceptance, 44, 251
Cold War, 226
Collingwood, Robin, 284
communication, 82, 114–15, 122–23
community, 185; in acquisition of knowledge, 167nn7, 8; examining assumptions behind research, 133–36; objectivity and, 148–52, 158–60, 166n4, 177–78; religious, 182; scientific, 95–96, 117–18
computability, of theories, 3
conceptual schemes, 181
conflicts of interest, of corporations, 7
consistency, as value of science, 58, 63, 66
constructivism, 173–74
contexts of relevance, 119
contextualized causal relations, 279–80, 283–86, 289, 291
corporations, 7. *See also* industry
counterfactuals, 174–75, 178
Creationism, 66. *See also* evolution
cultural evolution, 120, 207–11, 214n18

Damasio, Antonio, 154
Darwin, Charles, 107n4, 212
data. *See* evidence
Dawkins, Richard, 202–3
demonstration, of knowledge, 4, 86–87
Descartes' Error: Emotion, Reason, and the Human Brain (Damasio), 154
discovery, 4; social influence on claims of, 114–15; *vs.* justification, 16, 20, 22
Douglas, Heather, 54; on boundary between science and politics, 245–46; on inductive risks, 249, 253–54; on role of nonepistemic values, 70–75, 79
DuBridge, Lee, 224
Duhem, Pierre, 131
Duncan, Francis, 224–25

economy, 40–41, 275–76
Edinburgh school, 95
Einstein, Albert, 67–68
Eisenhower, Dwight, 225
empirical knowledge, 96–97, 99
empirical methods, 132–33, 203–5
Endless Frontier, The (Bush), 224
endorsement, of theories, 46–47. *See also* judgment
epistemic values. *See* cognitive/epistemic values
epistemology, 21, 130, 145, 174, 176, 183, 283; philosophy of science as applied, 14–16, 20; in scientific practice, 278–79, 287; in theory appraisal, 17–18
error, 178; and inductive risk in believing, 70–75; statistical theory of, 20–21
ethics, 6, 263, 271n7; goals of, 257–61, 269. *See also* bioethics; values
ethics committees, 269–71
evidence, 28–29, 87, 90, 101; and assumptions of causality, 131–32; and choice between equivalent theories, 58, 60–61; influence of values on, 3–5; insufficient, 73–74; interpretation of, 64, 157–59; objectivity of, 176–77; supporting social objectivity, 146–47, 153; in theory appraisal, 66–67; and underdetermination, 248–49. *See also* theory–evidence relation
Evita model, 194–201, 213nn6, 11, 214n12
evolution, 66, 107n4, 206, 212, 301; cultural, 207–11, 214n18. *See also* adaptationism
exemplars, 101
Experience and Prediction (Reichenbach), 15
explanatory unification, in theory appraisal, 19, 63

fecundity/fruitfulness, of theories, 58–59, 80
Federal Advisory Committee Act (FACA) (1972), 229–30, 247, 251–52
Fehige, Christoph, 183–84
Fermi, Enrico, 224
Foot, Philippa, 163, 169nn30, 32
Fox Keller, Evelyn, 286
fruitfulness. *See* fecundity/fruitfulness
Fuller, Steve, 145, 169n25

Garwin, Richard, 228
Gaut, Berys, 169nn30, 32
General Advisory Committee (GAC) (of Atomic Energy Commission), 224–26

Gergen, Kenneth, 145
Gilpin, Robert, 226–27
Golding, William, 225
Gould, Stephen Jay, 145, 202–3, 205
government: agencies of, 230–32, 241n4; bioethics and policymaking by, 256–57, 267–68; interactions with science, 116–18; response to scientific advice, 228–29; science and policymaking by, 220–21, 223, 227–28, 236–37, 239–40, 246–48, 253, 268–69; support for science, 222–24, 241n2; use of scientific knowledge, 116–17, 120–22; values in policymaking by, 251–52. *See also* politics; science-government relations
Graham, Loren, 118
Grene, Marjorie, 145

Haack, Susan, 167n8
Habermas, J., 108n11, 114
Hardimon, Michael, 251–52
Harding, Sandra, 145, 155, 157, 166n3, 168n19
Harvey, William, 167n13
Haugeland, John, 9
Hempel, Carl, 74, 79, 250, 252
Herrick, Charles N., 236
Hewlett, Richard, 224–25
HIV/AIDS, 117–18
honesty, as value of science, 127
Hull, David, 166n4, 211–12
human dignity, in law, 302–4
Hursthouse, Rosalind, 169n32
hypotheses. *See* theories

ideological values. *See* nonepistemic values; social values
impartiality, 43, 45; and theory acceptance, 24, 44; as value of science, 24–25, 80, 128–29, 137, 139–40
induction: consilience-achieving, 17–18; problem of, 131
inductive risks, 70–75, 249–50, 253–54. *See also* risk assessment
industry: research sponsored by, 275–76, 292; use of scientific knowledge by, 116–17
inference guidelines, in risk analysis, 233–35, 241n8
inference to the best explanation, 69
instrumentalism, 64–65
instrumentalization, 116, 302–3
intersubjectivity, 57–58, 65, 95–96, 114, 177, 182
intransitivity of sameness, 97–100, 102–3, 105

Jamieson, Dale, 236
Jones, W. T., 253
judgment, scientific, 2, 5–7; in legitimation of research applications, 46–47; norms of, 9, 249; in risk analysis inference guidelines, 234–35; uncertainty in, 73–74, 118–19; values and, 25, 252
judgments, value: influences on, 29; types of, 27; *vs.* value-assessing statements, 28
justification, 19; discovery *vs.*, 16, 20; reconfiguration of, 133–35, 148–49; values in, 53–54, 81

Kaplan, David, 176
Kelsen, Hans, 298
Kilgore, Senator, 240n1
Kissinger, Henry, 228
Kitcher, Philip, 48n9, 276–77
knowledge: acquisition of, 82, 154–55, 157–58, 167nn7, 8; application of, 98–103; in applied *vs.* pure science, 283–84, 286–87; claims of, 114–15, 147; distribution of, 114–16, 120, 124–25; evaluation of, 105, 112; influence of values on, 114, 149–50; models of scientific, 281–82; political uses of scientific, 116–17, 120–22; in popular media, 115–16; scientific as monist *vs.* pluralist, 130–31, 135–38; scientific *vs.* other, 112–13, 122–24; as social, 131, 136; social influence on, 81–87, 114–15; sociology of, 112; sources of, 67, 152–55, 168n15; types of, 12, 64, 262; and uncertainty, 73–74, 118–19, 121; from use-oriented research, 289–90; uses of scientific, 44–45, 64, 113, 118–19
knowledge, theory of, 14, 20. *See also* causality
Kripke, Saul, 176, 180
Kuhn, Thomas, 59, 80, 83, 95–96, 101, 249

Lacey, Hugh, 79–80
Lakatos, Imre, 20
Langton, Chris, 191
Laudan, Larry, 24, 64–65, 76n13, 79–80, 82–83, 168n14
law: causality in, 304–5; and legal method, 305–7; natural, 301; naturalism and, 295–96, 299, 301; philosophy of, 302; science's relations with, 294, 296–97; values and norms in, 297–98
learning, as social, 81–83, 86
Lenski, Richard, 200
Lewontin, Richard, 202–3, 205
life: artificial, 191; as source of values, 183–85

life sciences, 284, 289–90. *See also* bioethics
logic, in moral argumentation, 265
logical empiricism, 8–9, 295
Longino, Helen E., 48n9, 74, 83, 86, 148, 167n9; on motives for science, 87–88; on objectivity, 144–47, 157–60; on values, 80–81
Lorenzen, Paul, 257
Luhmann, Niklas, 112, 114, 124
Lysenko, Trofim Denisovich, 159–60, 169n24

Malthus, Thomas, 107n4
materialist metaphysics, 35, 40–41
materialist strategies, for research, 25–26, 34–38
Mbeki, Thabo, 117–18
measurement, validity of, 88
media, 120, 122–24
memory, as social, 83–84
Merton, Robert, 80, 94–95, 114
methodological rules, 27, 34–36, 40, 76n13; influence of values on, 43, 64
methodology, 129, 287, 297; autonomy of, 24, 43, 46; influences on, 64, 90, 278–81, 292; influences on choice of, 90, 138, 254n1; virtues of, 290–91
military, 224, 227–28
mind: facts independent of, 68, 169n26; sociogenesis of, 81–82; use of, 152; and objectivity, 152–53, 155–56
moral argumentation, 298; criteria for, 263, 269, 272nn14, 22; goals of, 260–62, 265–67; rules of, 264–65
moral judgment, and values, 300–1
Morgenbesser, Sidney, 14

National Academy of Science, 221–22
National Research Council, 221–22, 231, 240; on risk analysis, 232–34, 241n7, 254
National Science Foundation, 224, 228, 240n1
natural sciences: purpose of, 67–68; value freedom of, 128–29; *vs.* social sciences, 56
naturalism, 93–94, 96, 104–7, 169n32, 295; causality and, 304–5; law and, 297, 299, 301–7; science and, 294–95
nature, 10–11
Nelkin, Dorothy, 230
neutrality, 36, 48n9, 264; as value of science, 24–25, 128
neutrality-in-application, 25, 38, 43–45
Nixon, Richard, 227–29, 247

noncognitivism, and logical empiricism, 8
nonepistemic values, 4, 53; in application of knowledge to policy making, 70–75; as danger to science, 65–66; in determining aims of science, 64, 67, 79–80; epistemic and, 60, 74–75; in inductive risks, 54, 253–54; in problem selection, 55–58; in scientific disagreements, 248–49, 252–53; selection among, 71–72; in theory interpretation, 64–65, 75; in theory selection, 53–54, 73. *See also* cognitive/epistemic values; social values; values
normativity, 11, 103, 105–6
norms, 9–10, 261; application of, 108n11, 262; in bioethics, 258; in law, 297–98, 301, 306–7; of science policymaking, 251–52; of scientific practice, 95, 249; and use of knowledge, 84–85

objectification of values, 192, 212
objectivity, 9, 12, 134; of beliefs, 176–79; epistemic values and, 7–8; epistemological, 150, 152–55, 158; lack of, 181–82, 230; limitations of, 153–55, 168n23; metaphysical, 150–52, 160; not synonymous with universality, 164–65; of objects, 172–74; ontology of, 175–76; possibility of, 157, 168n18; of properties, 174–75; social, 150–52, 154, 158–59, 165–66, 168n14; social dimension of, 88, 95, 144–50; standards for, 148–49, 156, 158–59; through consensus of communities, 148–52; as value, 76n6, 127; of values, 160–65, 175, 182–86, 186; values not compromising, 57, 143
objects, 10, 38, 173; conceptualizations of, 9, 11–12; control of, 25–26, 32–33, 38–39; externality and continuity of, 172–73; objectivity of, 172–74; recalcitrance of, 10–12
Office of Defense Mobilization (ODM), 225
Office of Scientific Research and Development (OSRD), 223
Olby, R., 109n16
ontology: epistemology and, 174; and objectivity, 173, 175–76; in two-dimensional semantics, 176
Oppenheimer, Robert, 225

Packard, Norman, 192, 213n4
Pascal, Blaise, 60, 74
patents, in evolution of technology, 207–11
Path to the Double Helix, The (Olby), 109n16
peer review, 115, 148

perception, 11–12, 173, 175
Perl, Martin, 229
philosophical naturalism. *See* naturalism
philosophy, 96, 180; causality in, 304–5; of law, 302–5; naturalism in, 295–96; on science, 92–93, 105–6; on teleology, 191–92. *See also* science, philosophy of
pluralism: and objectivity, 144–45; in relation of science and values, 137–39; in scientific knowledge, 130–31, 135–36
"Political Impact of Technical Expertise, The" (Nelkin), 230
politics, 120, 298; bioethics in, 256–57, 267; boundary with science, 220–21, 231–34, 236–38, 245–47; distance between science and, 121–22; of ethics committees, 269–70; and government response to scientific advice, 228–29; risk management and, 232–34, 241n9; science policymaking and, 251–52, 270; of scientists, 226–27, 229–30, 247–48; use of scientific knowledge in, 113, 116–17, 120–22; values in, 240, 251–52. *See also* government
Popper, K., 112, 114
Presidential Science Advisory Committee (PSAC), 225–29, 247
Primack, Joel, 229
probability axioms, 179–80
problem selection, 52–53; and drive for application, 278–79, 292; influences on, 35–36, 90; values in, 1, 35, 46, 55–58, 137–38
properties, objectivity of, 174–75
Putnam, Hilary, 176, 178–79, 187n14

Quine, Willard V. O., 183

Rabi, I. I., 225
Radbruch, Gustav, 298
Rand, Ayn, 155, 158, 163
rational, the, 94. *See also* cognitive, the
rational reconstruction, 15–16, 19–20
rationality, 8–9, 128, 140n2, 179; and objectivity, 153–54
realism, 15, 179; monist *vs.* pluralist, 130–31; scientific, 64–65, 67–69
reality: mandates and choices of, 163–64; as mind-independent, 152, 167n11; objectivity and, 155–56; *vs.* beliefs, 147–49, 181–82; *vs.* social objectivity, 150–52
reconstruction, rational. *See* rational reconstruction
Reichenbach, Hans, 15–16, 19–20, 22
Reichenbach axiom, 179

reification, 102, 109n12
relativism, 145, 149, 166, 181
reliability, 88, 113; and inductive risks, 71, 74
religion, 64–66, 68, 182, 258
research: applied *vs.* pure, 275–76, 278–83; assumptions behind, 133–36; definitions of, 101; funding for, 241n2; influences on, 7, 114–16; motivations for, 21, 138–39; real-world *vs.* laboratory experiments in, 121, 279–80, 287; use-oriented, 25, 289–90; value of, 224, 289; values in, 4, 7, 21, 39–41. *See also* applications, of research; problem selection; science; scientific practice
Research and Development Board, 224
research papers, sections of, 4–5, 237–38
research strategies, 48nn6, 8; agroecological, 38; effects of choice of, 44, 46; materialist, 25–26, 34–38; variety of, 40–41, 43, 45, 49n16
results. *See* applications, of research
reticulate model, 64
risk analysis, 241n8, 254; in application of research, 41–42, 119, 121–22; and inductive risks, 249–50; National Research Council on, 240, 241n7; in science advising policy, 232–35; separation within, 241nn7, 9. *See also* inductive risks
Risk Assessment in the Federal Government (National Research Council), 232–34
Rodriguez, Eric, 6
role obligations, in science policymaking, 251–54
Roosevelt, Franklin D., 223
Ruckelshaus, William, 241n9
Rudner, Richard, 242n10

Salmon, Wesley, 88
"saving the phenomena," 17
Schweber, Silvan, 281–82
Schwemmer, Oswald, 257
science: aims of, 25–26, 53–55, 63–64, 67, 69, 148, 168n14, 276–77; as applied knowledge, 100–101; applied *vs.* pure, 276, 278–83, 291–92; authority of, 113, 116–17, 119, 122–24, 226–27, 231; boundaries of, 93–94, 103, 108n5, 112–13, 125; and the boundary with politics, 220–21, 231–34, 236–38, 245–47; dissenting opinions in, 117–18, 165–66, 167n5; distance between politics and, 121–22; ethics as, 263, 267; evaluation of, 105, 190; experts *vs.* lobbyists in, 270; good or bad, 7, 9, 71, 78–79; as idealized, 92–93; law's relations with, 294,

science (cont.)
  296–97; locations of judgments in, 238–39, 254n1; moments of activity in, 26, 44–46; motives for, 87–88; pluralism in, 137–39, 259–60; public discourse on, 114–16, 122–24; as social activity, 92–93, 96, 105–6, 112; social influences on, 90–96, 98–99, 103–5, 108n8, 128; success of, 105–6, 113; technology's relation to, 94, 275–78, 285–91; uncertainties of, 71–72, 118–19, 121, 232, 234, 236; use of term, 263; as value-free, 25, 129, 223, 225–27, 230. *See also* knowledge, scientific; research; theories; values; values, of science
science, history of, 90, 94–95, 104–5
science, philosophy of, 14, 106–7, 169n25; as applied epistemology, 14–16, 20; ontology of objectivity in, 175–76; underdetermination in, 131–33
science, sociology of, 95, 108nn7, 8, 133
Science Advisory Board, 222
Science Advisory Committee (to Truman), 225
"Science and Human Values" (Hempel), 79
Science as Social Knowledge (Longino), 149
science wars, 127–28
science-government relations, 240n1; after WWII, 224–27; effects of WWII on, 222–23; values in, 245–46. *See also* government
scientific practice, 88, 242n11; aims and ideals of, 30–31; government interference in, 117–18; ideals of, 30–31; intransitivity of sameness in, 97–100, 102–3; norms of, 9–10; values in, 26, 31–32, 237–40, 252; values of, 3–4. *See also* methodology; research
Scientific Revolution, 277–78
scientists, 99, 109n14, 124, 237; advising government, 225–26, 230; disagreements among, 248–49, 252–53; motivations of, 55–56, 138; mutual respect among, 95–96; participating in political issues, 229–30, 247–48. *See also* peer review
scope, as value of science, 58, 63, 66, 80; in applied vs. pure science, 283–84, 291; in theory evaluation, 18, 20, 22
Seaborg, Glenn, 224
Sellars–Putnam rule, 17
simplicity, as value of science, 58–59, 63, 80
Skolnikoff, Eugene, 230
Skusa, Andre, 207, 210
Smith, Tara, 182–84
social, the, 29, 79, 131; approaches to science, 127–28; definitions of, 140n3; as dimension of science, 128; influence on science, 90–96, 98–99, 103–5, 108n8; influence on scientific knowledge, 81–87; influence on tentative truth claims, 114–15; objectivity as, 88, 144–50; and relation to the cognitive, 22, 87, 129–30; vs. the rational, 94. *See also* values
social sciences, vs. natural sciences, 56
social values: influence on science, 137, 220, 240; and legitimation of research applications, 38, 41, 44, 46, 116–17; relations among, 30; in scientific practices, 32, 43–46, 237; vs. ideological, 53. *See also* nonepistemic values; values
sociobiology, 299–300
sociogenesis of mind, 81–82
sociology, 112; naturalistic approach in, 106–7; and science, 90–93. *See also* science, sociology of
"Spandrals of San Marco and the Panglossian Paradigm, The" (Gould and Lewontin), 202–3
Stalnaker, Robert, 176
statistics, 88; evaluations in, 2–3, 20–21; significance in, 70–75
stem cell research. *See* bioethics
Strohman, Richard, 286
subjectivity, 73–74; of values, 183, 185. *See also* intersubjectivity; objectivity
Super-Sonic Transport (SST), 228, 246–48
survival: as source of values, 183, 185; values supporting, 162–64, 169n34

technology, 39, 121; evolution of, 207–11; independence from science, 285–86; science's relation to, 94, 275–78, 281, 287–91. *See also* applications, of research
teleology, 191–93, 205, 305–6. *See also* utility
theories: adaptative hypotheses, 202–3; assertability of, 253; constraints on, 36, 40; definitions of, 30–31; effects of use-oriented research on, 278–80, 289–90; empirically equivalent, 61–63, 132; inductive risk in believing, 20–21, 70–75; interpretations of, 64–65; plurality of, 259–60; relation to applications, 281–82; standards for, 3, 59, 168n14, 178–79; structure of, 282; testing of, 45–46; unified, vs. local models, 279–80, 287; value of, 289, 291. *See also* theory-evidence relation
theory acceptance, 33, 44, 69, 129, 137; alternatives to, 45–46, 62; and dissenting opinions, 117–18; and empirical equiva-

Index

lence, 53–54, 58–63; errors in, 20–21; gap between data and hypotheses, 131, 140n5; and inductive risk, 249–50; influences on, 53, 75; standards for, 24, 147, 290–91; values in, 52–54, 73, 79–80, 251; *vs.* endorsement, 46–47

theory appraisal, 16, 21, 30, 145; of adaptative hypotheses, 203–5; cognitive and social values in, 24, 38; of empirically-equivalent theories, 59–61, 63, 132; evidence *vs.* ideology in, 66–67; non-epistemic values in, 66, 73, 79; rules of, 17–18; social influence on, 114–15, 120–21; truth in, 18–19

theory-evidence relation, 53–55, 74, 76n1, 79–81, 88, 131–32

transgenics, 38, 49nn14, 15

Truman, Harry S., 225

truth, 18, 187n14; as aim of science, 53–54, 63–64, 69, 127; effect of beliefs on, 181–82; influences on claims of, 114–15; and objectivity of beliefs, 177–79; in theory evaluation, 17–21; types of, 135, 179–80; as ultimate cognitive value, 183

truth conditions, 85

two-dimensional semantics, 176

underdetermination, 58; evidence and, 248–50; problem of, 131–33, 148

understanding, as social value, 31–33

Understanding Risk (National Research Council), 240

universality: as value of science, 128, 137, 139–40; *vs.* objectivity, 164–65

utility, 1, 67; as social value, 32–33, 91; in theory appraisal, 60, 74. *See also* teleology

Valsiner, J., 83

value-assessing statements, 28

values: appropriate roles in science, 1, 26, 38, 47, 52, 79, 143, 252; benefits to science from, 57, 127; epistemic and non-epistemic, 52–53, 248, 251–52; function of, 161–62, 165; in good or bad science, 7, 78–79; in government-science relations, 221, 223, 225–27, 237–39, 245–46; and inductive risk, 249–50; influence of, 2–5, 56, 128, 143, 234–39; lack of, equated with rigor, 129, 139–40; lack of truth or falseness in, 8; in law, 297–98; manifestations of, 26–30; moral judgments and, 300–301; as motivation, 21, 137–39, 149–50; objectification of, 192, 212; objectivity of, 160–65, 175, 182–86; pluralism of, 57, 137–39; in risk analysis, 232–36; science of, 190–91; and scientific knowledge, 113–14, 120; in scientific practice, 3–4, 242n11, 252; self-confirming, 184–85; social and cognitive, 26, 28–29, 38, 42–43, 80–81; social compared to cognitive, 24–25, 47, 49n18; types of, 2–3, 19, 78, 169n30. *See also* cognitive/epistemic values; non-epistemic values; social values; values, of science

values, of science, 53, 82; accuracy, 58, 66, 76n4, 127; applied *vs.* pure, 292; autonomy as, 24–25; consistency, 58, 66; epistemic, 275–76; fecundity, 58, 80; honesty, 127; impartiality as, 24–25, 80, 128–29, 137, 139–40; scope, 58, 66, 80; simplicity, 58, 80; truth, 127; universality, 128, 137, 139–140; *vs.* popular criteria, 123–24. *See also* objectivity

van der Veer, R., 83

van Fraassen, Bas, 19, 69

Vienna (and Berlin) Circle of Logical Empiricism, 295

virtues, 18–20

von Hippel, Frank, 229

Wallace, James D., 169n32

Weber, Max, 56

Whewell, William, 18

Wood, Robert, 226

World War II, 222–23

Wuketits, Franz, 299